Streaming Wars

Streaming Wars

How getting everything we wanted changed entertainment forever

Charlotte Henry

> **Publisher's note**
> Every possible effort has been made to ensure that the information contained in this book is accurate at the time of going to press, and the publishers and authors cannot accept responsibility for any errors or omissions, however caused. No responsibility for loss or damage occasioned to any person acting, or refraining from action, as a result of the material in this publication can be accepted by the editor, the publisher or the author.

First published in Great Britain and the United States in 2025

All rights reserved. No part of this publication may be reproduced, stored or transmitted by any means without prior written permission from Kogan Page, except as permitted under applicable copyright laws.

Kogan Page
Kogan Page Ltd, 2nd Floor, 45 Gee Street, London EC1V 3RS, United Kingdom
Kogan Page Inc, 8 W 38th Street, Suite 902, New York, NY 10018, USA
www.koganpage.com

EU Representative (GPSR)
Authorised Rep Compliance Ltd, Ground Floor, 71 Baggot Street Lower, Dublin D02 P593, Ireland
www.arccompliance.com

Kogan Page books are printed on paper from sustainable forests.

© Charlotte Henry 2026

The moral rights of the author have been asserted in accordance with the Copyright, Designs and Patents Act 1988.

ISBNs
Hardback   978 1 3986 2255 5
Paperback   978 1 3986 2254 8
Ebook   978 1 3986 2256 2

British Library Cataloguing-in-Publication Data
A CIP record for this book is available from the British Library.

Library of Congress Control Number
2025942907

Typeset by Hong Kong FIVE Workshop, Hong Kong
Print production managed by Jellyfish
Printed and bound by CPI Group (UK) Ltd, Croydon CR0 4YY

*For my n-squad:
Leah, Mila, Harry and Sadie*

CONTENTS

Preface xi
Acknowledgements xiii

Introduction: A media upset – 'The thing doesn't write itself' 1

01   See what's next: How Netflix changed everything 13
Enter the adverts 22
Netflix knows what you want 26
Going live 29
No news is good news 34
Netflix is a Joke 35
Company culture and the keeper test 37
Game on 40
Disruption 41

02   The arrival rivals: How tech became media and media became tech wars 43
Amazon 44
Apple 50
YouTube 55
Tech becomes media 60
Disney magic 62
Strategic alliance 64
Speeding up FAST 67

Media becomes tech 69
Death of the cinema? 73

03   Streaming sports: A good deal for fans or a sign of things to come? 79
Why sport matters to streamers 81
A difficult game to play 86
Netflix starts throwing punches 88
In DAZN 89
A fair deal for fans? 92
Bending the rules 99
Bin the black-out 101
Wrong Venu, wrong time 103

04   Livestreaming: YouTube, Twitch and a darker side to the wars 107
The top players 108
Twitch vs YouTube 111
Burnout 114
Is this thing on? 117
Zooming through the pandemic 118
The dark side of livestreaming 120
Terror, livestreamed 122
Making a connection 125

05   Audio battles, or, the beauty of music at our fingertips 127
Apple rescues the record industry 129
Spotify 133
Audible 137
Other players 138

Artists being paid 141
Nostalgia 143
The power of podcasts 145
Radio, someone still loves you 148
Direct connection 150

06 Streaming around the globe: Are we all watching the same stories? 151

India – they don't like cricket, they love it 151
China – in a world of its own 154
Into, and out of, Africa 157
Australia, New Zealand and Canada 159
Streaming a shared culture 162

07 Who pays the price? Do customers get what they deserve? 165

All over the place 165
Speciality viewing 167
Soaring subscription costs 168
The return of the advertisers 170
Send them to the kids' club 172

08 Reacting to a new reality: But what to do about AI? 175

AI and its impact 176
AI and audio 181
Disruption beyond AI 186
Filling the gaps 187
The rise of binge watching 188

Conclusion: We got everything we thought we wanted, but who is it good for? 191

My favourite TV shows to stream 199
My favourite podcasts to listen to 202
Notes 205
Index 237

PREFACE

The first time I became aware of streaming it wasn't called that and it only loosely resembled anything that we would call streaming today. I was watching video podcasts, themselves a pretty new innovation at the time. The titles I watched included *Tech News Today* on Leo Laporte's TWiT network and *What's Trending* from Shira Lazar – tech-focused shows that I could download to an iPad or laptop and watch on demand. It was quite thrilling to be on a family holiday and not have to watch the same as my parents and siblings!

Later came watching Netflix in the UK and the intrigue of settling down to watch breakthrough series *Orange is the New Black* and *House of Cards*. A total game changer.

Always one to test out new tech, I was excited by these arrivals. I can't pretend I predicted where it would all lead though. I'm (just about!) old enough to have grown up with a set number of TV channels. You watched what was on, when it was on. If you wanted to watch something 'on demand' you had to hope that the VHS video recording system you had would work. This sounds like the dark ages, but it wasn't all that long ago, I promise! The generation that followed mine will never know anything different than having a plethora of on-demand media at their fingertips.

The ramifications of that are enormous and a large part of the reason I was so keen to write *Streaming Wars*.

Because streaming is about more than the technology, the platforms and the shows. It's about more than Hollywood and studios and businesses. Streaming is about culture. Through services like Netflix, Prime Video and Peacock we get the TV shows, movies and sport that inspire fandoms, shape our opinions and form the basis of our conversations. Through Spotify and Apple Music we get the songs that are the soundtrack to love, loss and gym sessions. Podcasts and livestreaming allow us to hear from people from all over the world.

At the same time, because we have so many options, our culture has never been more fragmented. With the notable exception of live sport, we do not all watch the same thing at the same time anymore. Your best friend or close family members may never even have heard of the show you just obsessively binge-watched.

Having covered the tech and media industries for various outlets for a number of years, it became increasingly clear that the idea of differentiating between what is a tech company and what is a media company had become obsolete. Media companies have to bring their work directly to consumers through collaboration with big Tech or their own technology platforms. Tech companies are launching streaming services and winning major Hollywood awards.

As consumers we have never had more choice and more high-quality content on offer. Creators of all kinds have never had more opportunities to get their work out. At the same time, we are all having to pay for more and more services which keep putting up their prices. All the while, a significant number of artists are struggling to get by.

It is this plethora of contradictions that sits at the heart of the streaming wars and makes them so fascinating.

ACKNOWLEDGEMENTS

This book has my name on the front, but there are a huge number of people who supported me in turning it into reality and whom I cannot thank enough. Inevitably, I have forgotten some of you here, but you know who you are.

A huge thank you to my agent Lisa Moylett and her brilliant colleagues at CMM, Zoë Apostolides and Elena Langtry. Your wisdom and support were invaluable. Similarly, enormous thanks to my editor, Chris Cudmore, at Kogan Page. You helped improve *Streaming Wars* every step of the way and showed endless patience with me. To the interviewees and sources I spoke to, I appreciate you giving up your time for me and I'm thrilled to share your fascinating insights here.

Central to getting this book done was the support of my friends and family, whose tolerance of my combination of excitement at this project and panic as I worked through it made it all possible. Thank you Dan and Karin for your encouragement (and for keeping the secret I was working on this) right from the start. Nick and (another) Dan, those nights at heavy metal gigs were just the release I needed. Emma and my new friend Karin (this one pronounced Karen) who both know the joy and trauma of creative writing endeavours, your confidence in me meant so much. Anna, thank you for sharing your writing spot with me. It became the only place I could make real progress, but I won't tell anyone about it here! Annabel, your endless positivity is always a boost. Joel, thank you for your spreadsheet advice, which helped keep me on track. And to the rest of the Gnoming crew – your support means so much.

To my mum, Sue, my father, Paul, sisters, Gemma and Sarah, brothers-in-law, James and Marc, I love you all more than I can say. Whether it is through food, shouting at our stupid football team or having a cuddle, our family has a very special bond and I would never have been able to do any of this without you. Ma Ma, you're amazing and I love telling you all about what I'm getting up to and what shows you should stream next! I wish Pa, Grandma Rita and Grandpa Sam were hear to read this.

My darling nieces and nephews, you bring me more joy than I ever thought possible and helped me discover levels of love I didn't know I possessed. I can't wait for you to tell me how everything I've written about here is changing as you grow up. Aunty Char adores you.

Introduction: A media upset
The thing doesn't write itself

Hollywood on strike

Crowds swarm outside many of the major studio lots across Hollywood. Instead of the usual fans looking for selfies, it is the writers who are out on the streets. It is May 2023 and just a few weeks later they will be joined by the actors who perform the material they write. These artists will also spend the summer outside in the baking heat instead of creating and performing on the other side of the gates, as the unions representing writers, actors and the studios fail to come to an agreement for a new contract.

During the strikes, union members waved placards emblazoned with slogans like 'The CEOs have yachts. Writers have mortgages' and 'Turns out, the thing doesn't write itself'.[1] This standoff was the result of tensions boiling over, tensions that had been exacerbated by the shift to streaming. And boiling it was. At one point there were allegations that Universal, one of the major studios at the heart of the dispute, had pruned some trees outside its studios to remove a source of shade from the picketers in a bid to try and get them into an air-conditioned room with a negotiating table.²

The writers and actors were reacting to a new reality, one they were not sure was in their best interests. The type of work they were being asked to make, and the way they were being asked to make it, had changed radically in a relatively short space of time. Ever more sophisticated artificial intelligence (AI) tools are only going to cause further disruption. The picketers wanted to draw a line in the sand.

Younger readers may find it hard to imagine a time before streaming. A time when there was not endless 'content' instantly available at our fingertips. Services like Netflix, Disney+ and Apple TV+ have fundamentally altered how television shows and films are made, distributed and consumed. Tech companies have become media companies and vice versa. The likes of Apple and Amazon now produce original television shows and movies as well as selling widgets, while Disney and Paramount have had to build tech platforms on which to distribute their work.

As consumers, we have become accustomed to browsing through an almost limitless menu of entertainment options, including work produced by individual creators on social media platforms – another challenger to the giant studios. In January 2023, research from Kantar found that in Great Britain alone 16.24 million households subscribed to at least one streaming service.³

The business and cultural changes this has brought about are profound, and the way that these services have grown has been truly staggering. In general, people are becoming more and more comfortable with renting, not owning, things. This is particularly true when it comes to media. People feel that there is no need to permanently have the DVD of a favourite movie or the box set of a favourite show taking up space and collecting dust when they are available at a click of a button. We no longer own many of the items with which we used to demonstrate our personality. Instead, we rent them via a monthly payment. An extension of this culture is services for delivering meal kits, with ingredients for meals already portioned out, and others for renting clothes.

Yet consider this: Netflix, generally regarded as the 'original streamer', only actually started offering an online catalogue in 2007 and it was only available in the US at that point. At the time, the company's financial statements barely mentioned the new 'online video' service, reflecting how relatively unimportant it was to them overall. The focus remained on sending out DVDs for rental via the post, a service it only stopped providing in September 2023. The pace of disruption has then, in relative terms, been fast.

Back to the hot streets of Hollywood. The fundamental shift in the entertainment industry might sound great for those who make TV and movies, the people who are on strike and protesting. And in many ways it has proved to be so. Actors, directors and producers have had more opportunities than ever to get their work in front of an audience. Some are handsomely rewarded. Streamers have proved more willing to back riskier, edgier projects. Yet, as those stickers and signs demonstrated, it is not that simple.

AI concerns

The Hollywood industrial action ended up lasting for many months. The writers were on strike for 148 days. The actors lined up with them on 14 July 2023 and did not return to work until 9 November. The impact on release schedules was huge, with movies and TV series pushed backed or even cancelled.

Both groups had a range of concerns that led them to the picket line, with some of the practices used by streamers high on their list of complaints. These included a lack of transparency from the companies, which actors and writers felt limited the amount they could earn in things like residual payments, and smaller writers' rooms, i.e. fewer people getting work in the first place. In addition, there were worries that even though streamers were making series shorter, the writers had contracts that locked them in for the same amount of time as before, meaning they could not go and get other work, limiting their earnings.

Looming over all of this were concerns about how AI could be used to produce content, with the contract row erupting just at the time that AI services like ChatGPT, Dall-E and Midjourney got mainstream recognition. The fears from the television and filmmaking workforce encompassed everything from the use of AI tools for generating ideas and scripts to using the images and voices of actors, reducing their value.

There is little doubt that, as time goes on, writers and actors will be banging on the doors of studio execs again and again to discuss these topics. Mere months after

Hollywood resumed business, OpenAI released a product called Sora that allows a text input to produce a short video. Given this is only the start of things, you can see why those in the creative industries have very real worries.

How did we get here?

There are many elements to the streaming story that have led us to this point. These services were initially considered outliers, even inferior to the 'proper television' we were used to. They mostly began by offering shows and movies that traditional broadcasters were not all that interested in or that had been long forgotten. The insurgents hoovered it all up in a desperate bid to build a catalogue.

The moment that altered the industry most profoundly was when Netflix started producing work of its own. It did not happen overnight, but mega-hits like *House of Cards* and *Orange is the New Black* undoubtedly took Netflix mainstream. Suddenly, it was offering must-see TV – series that people were talking about with friends, family and colleagues. Shows you didn't want to miss the latest episode of.

Netflix splashing the cash on *House of Cards* inevitably caused waves in Hollywood. Reports from the time reveal how taken aback the industry was at this brand-new development, with Netflix outbidding major networks like HBO and AMC and committing to a mammoth 26 episodes across two seasons. 'Because it's never happened before' is one key phrase that jumps out of a very short article in New York's culture website Vulture, which goes on to effusively describe the deal as a 'landmark move away from networks and towards a new internet-derived future'.[4] There is almost a sense of shock that a mere streaming company could be buying something involving such big names, which reflects industry opinion at the time.

The show, as we now know, proved to be hugely successful prior to Spacey facing very serious sexual offence allegations and it coming to an end. *House of Cards* boosted Netflix as a provider of content, proving it could produce must-see TV of the kind usually seen on HBO and the BBC. It undoubtedly paved the way for others to enter the market, triggering the start of the streaming wars.

As time went on, every media company realized that they needed a direct-to-consumer (DTC) option, a platform that they owned, offering the work that they had commissioned and made – and spent a fortune making. The story of streaming's growth is not only about content, though. As we know, every action triggers an equal and opposite reaction. So tech firms with deep pockets like Amazon and Apple, once seen as the geeky opposites to the creativity and glamour of Hollywood, decided they wanted a slice of the pie too, shifting the dynamics once again.

Technological advances, most notably super-speedy broadband and ever-cheaper smart TVs, contributed to the widespread uptake of such services. The former meant that content could be delivered quickly and in high quality. The latter allowed it to be consumed easily from an app on a television set. No more being hunched over a laptop or plugging in various cables into a display. Furthermore, improved mobile technology such as better phone screens and 4G then 5G cellular networks meant that consumers could even take in the latest episode of their favourite show on the go. If a billion people have an iPhone, it makes sense that Apple wants to make some of what those customers watch on their devices and charge those customers for it.

Pandemic booster shot

There is another significant factor that we need to consider. Many of us want to forget the difficult years of the Covid-19 pandemic, but the truth is that it was hugely beneficial to streaming services. It did adversely impact their ability to produce new content as crews and cast were unable to come together to shoot; however, with people having little else to do, entertainment providers saw a boom as demand accelerated and huge parts of the public had little else to spend their cash and time on.

On 27 March 2020, as the true seriousness of the situation was becoming apparent and lockdowns were being imposed around the globe, WarnerMedia published data regarding the effect on its DTC offering. The previous week in March had seen the entirety of television get a 20 per cent boost from the same time in February 2020. On Warner's own platform, then called HBO Now, it found that 'binge watching' was up by 65 per cent, with movie viewing spiking by 70 per cent.[5]

Consumers have certainly become used to 'binge watching' – watching every episode in a season in a short space of time. With that, the concept of everyone gathering around a television at the same time to watch the night's big episode and then discussing it at work the next day has rather disappeared. If such water cooler moments do still exist they are normally based around a sporting event, which can only really be consumed live and are still often shown on a mainstream linear television channel. Think Super Saturday at the London 2012 Olympics or the clash between Lionel Messi and Kylian Mbappé at the 2022 World Cup Final.

There are other non-sporting – usually news – events that can only be taken in at the time too. The attempted assassination of Donald Trump on 13 July 2024 happened on live TV, for instance. Many Brits will be able to tell you where they were when they watched the announcement of the death of Queen Elizabeth II. Away from sport and news there are still some big drama series that can get people to stop whatever they are doing – the whole of the UK seemed to watching the final episode of *Happy Valley* in 2023 and nobody appreciated a *Succession* spoiler. Popular reality TV series can have a similar effect, but it's not as it once was. How could it be?

Having previously been seen as a bit downmarket, streaming services now dominate the culture and win prestigious awards. Indeed, in 2022, Apple TV+ became the first streamer to pick up Hollywood's highest honours – the Best Picture Oscar for *Coda*. Although Apple had purchased the film – which focuses on the story of the only hearing member of a deaf family and her musical ambitions – from the Sundance film festival instead of it being a pure in-house original, it was a huge moment. (One that Netflix had been striving to have for years.) The pandemic was a defining moment in that shift, as all content went to streamers.

Looking to score

Streaming services are increasingly moving into the world of sport too. This includes Premier League football on Prime Video and a range of US sports on ESPN+. Google spent a reported $2 billion to get the rights to show certain NFL games on YouTube TV. Messi's decision to move to play football for MLS side Inter Miami prompted some interest in Apple TV+, which had struck a deal to stream every single MLS match for a decade prior to the Argentina superstar's arrival. Indeed, so crucial was his presence that it was reported that he got a cut of the subscription revenue generated by Apple's Season Pass offering.[6]

At the time of writing, key events like the NFL's Super Bowl and the football World Cup remain with traditional broadcasters, but there is plenty to suggest this could change in the future. Streaming companies are willing to offer big bucks for lucrative sports rights. This would be a huge shift and put even greater pressure on the likes of Sky Sports in the UK and cable providers in the US. For many, live sport is the only reason to continue paying those providers significant sums of money for their overall bundle.

In the UK, certain major sporting events, known as the Crown Jewels, must be shown on terrestrial television.

These include football's World Cup and FA Cup finals, the Wimbledon finals in tennis, and horse racing's The Derby and Grand National. This may not last forever.

The business of streaming

As well as changes to the type of content being offered, there are changes on the business side of streaming too. Crucially, as time has gone on, investors and top executives have increasingly begun to demand a focus on revenue growth, not just the growing subscriber numbers, which was the original goal. Companies have had to update their strategies accordingly. One of the key moves was the introduction of cheaper, ad-supported membership tiers from Netflix and Disney+ as they attempted to get more people to pay something, anything, instead of sharing an account with friends and family or passing up on the service altogether.

The challenge faced by all streaming providers is that as the number of services available increased, so did the frustration from viewers. Customers started to realize that the programming they wanted to watch was spread across a wide range of platforms, all of which required a separate annual or monthly payment. There is a limit to how many different providers people are willing to fork out for. The later arrivals to the market struggled to make an impact. Indeed, some quickly became part of a bundle offered in conjunction with linear television providers such as Sky, who offered some Peacock programming with their packages in the UK (the two broadcasters have the same parent company).

Whereas originally a service would release whole seasons of a show at the same time, allowing viewers to binge-watch, they are now spreading things out and releasing episodes on a weekly basis. For instance, Netflix split out the release of the fourth season of its hit series *Stranger Things* into two volumes – one in May 2022, the second in July 2022. This meant that to watch it all as soon as it came out fans had to have a paid subscription for multiple months and in at least two financial quarters, a boost to the company's results.

There is a distinct possibility that we could be returning to the kind of structure seen in the classic cable TV days, when the cost of a bundle was relatively high but consumers got a lot of what they wanted from a few key gatekeepers. And you had to wait a few days for the next episode of your favourite show! Maybe those water cooler moments have not completely disappeared after all.

While much of the attention is on visual content, the discussion around streaming should not forget the changes in music and other audio industries. When Steve Jobs finally managed to charm and bully the music industry into submission, artists were delighted with the arrival of the likes of iTunes. It seemed to solve the ongoing problem of piracy posed by Napster and other websites that offered free downloads of their music at varying quality. With iTunes (now Apple Music) and Spotify coming to prominence, it was easier to get high-quality music recordings legally than illegally. It has also never been simpler for musicians to publish music. However, even the biggest stars worry they are not being remunerated fairly for their work, an ongoing source of tension within the music industry.

Podcasts are historically associated with Apple and the iPod, but Spotify has, in recent years, invested considerable sums into the medium, launching originals and putting a paywall around some popular content. It had to rethink this approach amidst a difficult post-Covid pandemic market in 2023 and scaled back its efforts.

For some, streaming is another phenomenon altogether. It is not about big budget shows and movies but content produced live by online creators. This often revolves around video games, as people tune in to watch others play and pay money to support their favourites.

As ever, the story told in this book is not just about companies or content. There are key individuals behind these seismic shifts – the likes of Netflix co-founders Reed Hastings and Marc Randolph, Zack Van Amburg and Jamie Erlicht at Apple TV+, and Spotify's Daniel Ek. They, and many others, have played key roles in how the public consumes media.

Streaming services of all kinds are, to a lesser or greater extent, the centre of most of our media diets. It's time to really look at the fallout caused by the arrival of these services and explore the possibilities of what might come next.

The streaming wars have only just begun.

01
See what's next
How Netflix changed everything

Contrary to what we all think, Netflix was not actually the first company to offer on-demand digital media. You could buy videos from iTunes and the Amazon Unbox service (which eventually morphed into Prime Video). However, it was undoubtedly the first mainstream streaming service, as we have come to understand the concept.

In direct contrast to what other companies were offering at the time, Netflix customers didn't pay to own and watch an individual TV series, episode or film. Instead, they got access to its entire digital catalogue for a reasonable monthly price. There were no adverts, no hoping the movie you wanted was available at the rental store or worrying about returning a physical disc by a relatively tight deadline if it was. This was just uninterrupted viewing whenever you wanted it. Looking back on those early years, they seem almost idyllic.

It remains no exaggeration to say that Netflix altered the media landscape beyond recognition, leaving some of the most established names in the industry scrambling in its wake. The service tends to be top of the charts in terms of subscriber numbers, indicating that if someone pays for any streaming service they probably pay for Netflix. While 'Netflix it' (somewhat different to the more flirtatious 'Netflix and chill') has never *quite* became the same as 'Google it', it is not far off. When we ask if something is available on a streamer, by default we tend to mean whether it is on Netflix. The bottom line is, when we think of streaming, we think of Netflix.

It has taken years for many of the linear television networks to have anything resembling a viable direct-to-consumer (DTC) digital offering that can challenge Netflix. For far too long, those companies desperately tried to keep alive the traditional cable bundle, through which customers signed up to a package containing a wide variety of channels, most of which they never watched or even knew existed. All the while, the companies generated huge profits. Netflix was more than happy to help destroy that model.

Subscriptions are, of course, far from revolutionary in themselves. Newspapers and magazines have offered them for years. However, on the back of the streamer's success, the 'Netflix model' is being applied in a variety of sectors. Take meal kit boxes and online craft beer clubs, for instance. You pay on a monthly or annual basis and the thing you want becomes available to you, either immediately digitally or via physical delivery. Not all that long ago we would have found it utterly bizarre to have itemized dinner ingredients or fancy IPAs dropped at our front door on a regular basis. That is no longer the case, and it is in no small part down to Netflix.

This is great in lots of ways, most notably because it means we get media content more easily, delivered to us more easily and conveniently. In other ways, it's a bit like a frog being boiled, which only realizes the pot is hot when it is far too late. Every individual subscription is fairly reasonable, but if you take a step back you suddenly find that there is an awful lot of money heading out of your bank account each month. How many times have you said, 'Oh, it's just another £5/£10/£15 a month'? I know I have (although I always convince myself that signing up to the latest streaming platform is 'essential for work'). In large part we have Netflix to thank/blame for this too!

The service got us used to subscribing and comfortable with leasing media instead of owning it. For decades, displaying the CDs, DVDs and books told people something about you, but with the rise of iTunes, Spotify, Netflix and the like, this has rather disappeared. It is much harder to steal a surreptitious glance at a potential new partner's Netflix watchlist than it is their DVD collection. While we are watching more and telling Netflix more about ourselves, we are actually telling people less about our preferences and personality. Perhaps this is why we like to reveal ourselves in new, even more public, ways, such as carefully curated social media profiles.

Whatever the consequences, for a certain generation, the whole concept of owning the media you like, be that CDs, box sets or books, is an almost alien one. The rise of Netflix is a key part of why that is.

The Netflix story – Act One

Founded by entrepreneurs Reed Hastings and Marc Randolph in 1997, Netflix initially mailed DVDs for rental, a business that it did not shutter until September

2023, two and a half decades after it began. (For those wondering, the first DVD the company ever sent out was *Beetlejuice* starring Winona Ryder. The last was *True Grit* from the Coen Brothers.)[1]

Netflix's website launched in 1998, and in 1999 it began offering a subscription service whereby customers would receive unlimited DVDs and were not hit by the late fees and deadlines that traditional rental companies imposed. Not surprisingly, this proved popular and by 2003 it had a million members in the US. Three years later that number was five million.

This might indicate that its ascendancy was simple, especially as Netflix now seems something approaching untouchable, but things were not straightforward. Indeed, the company nearly didn't survive and tried to merge with a bitter rival in a desperate attempt to keep the lights on. Randolph, who left in 2003, recounted what happened in a post on LinkedIn in April 2023:

> We'd been struggling since 1998 to find a way to make it work, and by the summer of 2000, we were finally seeing light at the end of the tunnel. Our no-due-dates, no-late-fees subscription model was a hit. Customers were pouring in and the company was growing like crazy. Running a subscription business takes a lot of cash, since you pay acquisition costs up front while the revenue comes in over time. But it was the height of the internet boom. Cash was plentiful. Until suddenly it wasn't.[2]

So off he, Hastings and Chief Financial Officer Barry McCarthy went to the HQ of… Blockbuster, the traditional video-rental company whose stores dominated the movies-at-home market before the streaming wars started.

They were heading behind enemy lines. Netflix now needed the firm it had been determined to kill to help save it. Desperate times called for desperate measures.

Sitting in the home of their bitter rival, the Netflix execs proposed a deal that would see the companies merge, with Netflix running the online offering and Blockbuster running the physical stores. The Netflix team said their company was available for the price of $50 million. This was not exactly well received on the other side of the negotiating table. In that LinkedIn post, Randolph recalled that 'there was perfect silence. Their words were "We'll consider it", but we could tell they were fighting to suppress laughter.'

As we now know, the Netflix team more than turned things around and the tables rather turned in the years that have followed. As Randolph noted: 'The company that Blockbuster could have purchased in 2000 for $50 million has a market cap exceeding $150 billion. And that company with 9,000 stores? Now it's got just one.'

More than that, the Blockbuster corporate entity filed for bankruptcy in 2010. At the time of writing, that one store that Randolph referred to, a franchise unit in Bend, Oregon, remains open[3]. It is essentially a nostalgia-fuelled tourist destination where fans can purchase kitschy Blockbuster-themed merchandise. There is even the occasional wedding there. Very romantic.

The Netflix story – Act Two

The crucial moment that helped Netflix make it through those difficult times and get to where it is today came in 2007, when it introduced streaming in the US. This service was then expanded internationally, first to Canada in 2010 and then to the Caribbean and Latin America a year later. Netflix didn't arrive in Europe until 2012, when it became available to customers in the UK, Ireland and the Nordic nations. That year, the service had 25 million members, with growth and expansion continuing from there. By 2016 the streaming offering was global and nobody was worrying about Netflix not merging with Blockbuster… apart from maybe the Blockbuster execs who had sneered at the prospect all those years earlier.

Even more profound change was on its way. Not only was Netflix going to be a distributor of other people's content, but it was also going to become a creator of TV shows and movies too. If you're reading this book you've almost certainly watched Netflix original content. Lots of it.

Its first original was *Lilyhammer*, starring Steven Van Zandt (Bruce Springsteen's E Street Band guitarist) alongside Trond Fausa and Steinar Sagen.[4] The series tells the tale of a New York mobster who, having testified against his former partners in crime, was forced to flee and hide in snowy Lillehammer, Norway, a place perhaps best known for hosting the 1994 Winter Olympics. The show ran for three seasons between 2012 and 2014 and even succeeded in picking up some awards, albeit relatively low-profile ones.

It was not the biggest hit, but a Rubicon had been crossed. Netflix had shown it could make its own stuff and that it was pretty good. *Lilyhammer* was a clear indication that this distributor could become a creator. It seems an obvious move now, and in the years to come everyone else would follow suit. At the time, though, it was a gamble, and an expensive one.

The key shift came a year later in 2013, when Netflix put *House of Cards* out into the world, spending $4 million an episode and $3 million on development.[5] As discussed in the introduction, this release sent ripples throughout Hollywood. Starring Kevin Spacey and Robin Wright, it was a remake of the classic political drama first made in the UK by the BBC. The Americanized version of politician Frances Urqhuart, his wife and colleagues gripped viewers around the world. The series led to Netflix becoming the first internet service to pick up a Primetime Emmy (it won three that year). The show continued until allegations were made against Spacey and it was eventually wrapped up.

Orange is the New Black, a tale of love and loss in an American women's prison, was the next Netflix series to grab serious attention. The show won various awards from 2014 onwards, including Outstanding Guest Actress in a Comedy Series for Uzo Aduba following her haunting performance as Suzanne 'Crazy Eyes' Warren. Laverne Cox picked up the same award in 2020, following the show's final season for an equally powerful portrayal of transgender inmate Sophia Burset.

As well as being engaging, the shows were all high-quality – offering production values reminiscent of work put out by HBO and other high-end broadcasters. Having stars like Spacey on board was also a big public endorsement, showing that Netflix could recruit the best and mix it with the big boys of the entertainment industry.

They also had the freedom to explore topics that most mainstream broadcasters were reticent to. *Orange is the New Black* pulled few punches when looking at issues around gender dysphoria and transition, sexuality and the abuse of female inmates. Few mainstream linear broadcasters, beholden as they are to advertisers, would be likely to produce such challenging work.

By the second quarter of 2019, Netflix boasted more than 60.1 million paying subscribers in the US, with a further 1.6 million using the service on a trial basis, according to a research note at the time from Leichtman Research Group (LRG). The analysts calculated that, in total, half of US households had subscribed to Netflix by that point. Furthermore, 60 per cent of US households said they had access to Netflix.

However, even in an age of mass digital media consumption, Netflix has encountered issues. Most notably, password sharing – where users, normally across multiple households, access the same account – had been increasing. LRG's Emerging Video Services 2019 survey revealed that over a quarter (27 per cent) of adult Netflix subscribers agreed that 'My Netflix subscription is shared with others outside my household'. This figure had risen from 15 per cent in 2014. Originally, Netflix execs, and those from other streaming services, were pretty relaxed about password sharing. They accepted it was going to happen and almost saw it as a marketing opportunity, a way of getting their brand and content in front of more people. They hoped that at least some of those who were logging in with others' passwords might choose to pay up in the future, or at least recommend a show and encourage someone else to pay.

The LRG research noted that in Netflix's Q3 2016 earnings call, Hastings, then CEO, seemed unbothered about password sharing. He commented that his company had 'no plans on making any changes... Password sharing is something you have to learn to live with because there's so much legitimate password sharing, like you sharing with your spouse, with your kids...' Most telling of all, Hastings said: 'We're doing fine as it is.'[6]

And they were doing fine. Netflix continued to produce hit shows on a relatively consistent basis. Among the biggest was *Stranger Things*, a supernatural mystery, the first season of which was released in 2016. It proved so popular that, after appearing in the fourth season of the show, Kate Bush's 'Running up that Hill' and Metallica's 'Master of Puppets' were propelled back up the music charts decades after their original release, having been introduced to a new, younger audience in key scenes. There was also the Korean thriller *Squid Games*, *Inventing Anna* – based on the true story of alleged con artist Anna Delvey – and *The Queen's Gambit*, which prompted a huge take-up in chess playing.

But this positive trajectory did not continue uninterrupted. Users, under growing financial strain and with access to various alternatives, began giving up their subscriptions. Close to a million subscribers abandoned Netflix between April and July 2022. This marked the second successive quarter in which the service had lost subscribers.[7] Concerned, Netflix started taking the issue of password sharing seriously. If you were watching it, the company really wanted you paying for it. Perhaps it is no surprise that the level of concern escalated. Perhaps the company should have been paying more attention before.

In the first quarter of 2023, LRG research found that 27 per cent of all DTC subscriptions were being used in more than one household, with 13 per cent being 'used and paid for by those that also share them with someone outside the household'.[8]

The company's increased focus on password sharing was demonstrated in a shareholder letter in April 2022. The company revealed in the letter that 100 million users were sharing their passwords: 'Our relatively high household penetration – when including the large number of households sharing accounts – combined with competition, is creating revenue growth headwinds.'[9] That is corporate speak for 'too many people are using our service without paying and we've had enough of it'.

Enter the adverts

Part of the attempt to keep people both watching and paying was the launch of a cheaper ad-supported tier. This is despite the fact Netflix had long railed against the idea of having adverts on the platform. The 'Basic with Ads' package started rolling out on 3 November 2022. In the US, it cost $6.99 while it was £4.99 in the UK, considerably lower price points than any other plans the company or most of its rivals were offering. The first ad ever to run on Netflix appeared in Canada. It was from the jewellery firm Tiffany & Co and featured Beyoncé.[10]

Despite the presence of the pop icon, there was not huge initial uptake of the ad-supported tier, not least because many popular series were unavailable for the lower price.[11]

Furthermore, the whole thing felt like a culture clash or, worse, a broken promise. Netflix had always seemed to operate on the premise that if you pay you can watch whatever you want, undisturbed by ads. It was a huge leap for consumers to accept the return of advertising interruptions, a format that felt old fashioned and the antithesis of everything that Netflix had originally been created and seemed to stand for. It felt like a fundamental deal with the consumer had been destroyed.

Given Netflix's pioneering history and strong market position, its abandonment of its ad-free promise allowed its rivals to follow suit. It gave the likes of Amazon the space to introduce advertising in ways that were even more frustrating for us viewers. (With Amazon, the ads were added by default and you had to pay more to not have them.) If Netflix was doing it, why shouldn't they?

It is worth saying at this juncture that there is nothing wrong with adverts per se. Quite the opposite. They are perfectly appropriate on things like linear live TV, YouTube and podcasts. Sometimes they can flag an interesting, useful product to the viewer or listener and, if nothing else, it is what we are used to in such spaces. However, Netflix played a crucial role in creating our streaming environment and had seemingly made it clear it was doing so on a subscription-driven, ad-free basis. They were different, the upstart, and we viewers were paying a premium to not have the show we were watching interrupted or delayed by someone trying to sell us something. Introducing a reduced cost advertising tier shattered that illusion.

Netflix was successful in building out the ad-tech behind the new offering, something it initially did in partnership with Microsoft. However, it took a while for the ad business to really start scaling.[12] Perhaps because users simply did not associate Netflix with advertising, they were initially resistant and ad-tier take-up wasn't massive. It did, though, get to 70 million subscribers two years after its launch.[13] By May 2025, that number had risen to 94 million, with an average viewing time of 41 hours per month in the US.

Speaking to advertisers at the upfronts that month, the third time Netflix had had to host the ad industry staple, Amy Reinhard, Netflix's President of Advertising, explained that the service was watched by more 18–34 year olds, the crucial demographic for advertisers, than any cable or broadcast network in the US.

What set her company apart, Reinhard explained, is their 'ability to marry art and science, combining best-in-class technology with the shows and movies that everyone is talking about and watching'. She also said that 'while a lot of companies are either/or – either they have great technology, or they have great entertainment – our superpower has always been the fact that we have both.' It's the kind of talk that gets ad execs salivating.

Reinhard insisted that 'because our audience is unique, engaged and attentive, a dollar spent on Netflix is more valuable than a dollar spent anywhere else'.

In the same presentation, Reinhard announced that Netflix was rolling out its own ad-tech called the Netflix Ads Suite. Bringing this in-house really underlined the commitment that a streamer once so opposed to advertising now had for the revenue stream. The streamer deployed Lily Collins, in character as Emily from hit series *Emily in Paris*, to explain that brands could 'now target more than 100 interests in 17 categories – including life stages' with this new technology. Advertisers love advertising on digital platforms like streaming because of the granularity of the data they can get back.[14]

There had, though, been some evidence that subscribers watch less. Research released in September 2024 by Digital i, a streaming ratings firm, revealed that those on the more expensive plans watched 34 minutes, about 40 per cent, more Netflix each day.[15] We know from our linear TV experiences that the ad-breaks are a perfect spot to make a cup of tea, nip to the bathroom… or get distracted by something else. Netflix introduced those interruptions into our previously unbroken, binge-filled streaming viewing.

Like ads or not, there isn't much the consumer can do about their introduction. When it comes to Netflix, the choices are pay less and watch ads, pay more and don't watch ads, or simply don't watch the thing you want. As we will see, this all adds to the sense that we are going 'back to the future', in which our brave new world of streaming actually resembles the world of linear TV we thought we were leaving behind.

Bringing it all together

Netflix hinted at another change in direction in June 2025. It made a deal with TF1, the largest TV channel in France, to host both its live channels and on-demand content in the broadcaster's home country. The streamer had previously avoided doing anything to support its rivals, making the deal significant, albeit relatively small.

The arrangement with TF1 hints that Netflix might be interested in becoming not just a content maker and provider but an aggregator in a manner similar to Amazon Prime Video and Apple TV+, something to be discussed shortly. In a note unpacking the deal, Tom Harrington of Enders Analysis said that it 'could be seen as a proof of concept with less of the risk and complexity of one with a rival subscription operator'.[16] After the introduction of both advertising and live sport, we have all learnt never to dismiss the idea of any move Netflix might make in the future.

Netflix knows what you want

Netflix may seem to be ubiquitous, but our individual experiences of the platform are not all the same. The company drills into the data its collects so the way one person's home screen looks will be very different to another's. They will be recommended different shows, based on various criteria.

Factors considered include how a viewer interacts with Netflix, encompassing viewing history and the ratings they have given shows or movies, and what others perceived to have similar tastes also like. Various other details about a title are considered too, including everything from the genre it's in, the actors featured and its year of release.

Furthermore, Netflix looks at the time of the day someone uses the service, what language(s) they watch in, what devices they watch on and how long they viewed something for.[17]

Indeed, even if the algorithm recommends two viewers the same series or movie, the image used to promote it could be different – selected to maximize the chances of the viewer pressing Play.

Data collection lies at the heart of Netflix's operation, and its success. It's all about keeping you in the app for as long as possible and watching as many things as possible. In the company's 17 October 2024 earnings call, co-CEO Ted Sarandos commented that the company considers engagement 'our best proxy for member happiness because when people watch more, they stick around longer'. He added that customers then 'talk more about Netflix, which drives acquisition. And they place a higher value on their Netflix subscription.'

For customers, this determination to give us what we want, or at least what we think we what, can be great. It means we get recommended shows and movies that we are likely to enjoy because they are similar to things we have watched already. However, it does also mean you can get a bit stuck only watching the same kind of production, instead of discovering something truly new. Like American 30-minutes comedies such as *Friends* or *How I Met Your Mother*? Here's a bunch of others. Action movies more your thing? Netflix will gladly recommend dozens more. It all underlines how tech lies at the heart of what is ostensibly a media company.

Of course, Netflix is not averse to using its power to push certain movies or shows that it really wants to be seen. This tends to happen when it has spent a lot of money on something and wants to get a return on its investment.

Despite having all this technology, the company insists it values quality and creativity over all else and that its primary focus is making the best stuff possible. Somewhat bizarrely, those involved in making content for Netflix, along with its rivals, have been left frustrated by how little the company shares publicly. What actually counts as a hit? The company has become a little bit more transparent as time has gone by, but it is still hard to quantify what it really considers to be a success.

The introduction of advertising may well change this even more. Advertisers simply demand to see the numbers before they sign a cheque. Yet the creators of the content in which ads appear may still not know how their work is actually performing. While this may relieve them of the stress that constantly following the ratings brings, it also leaves them somewhat in the dark, wondering whether their show or movie is hitting the mark.[18]

Back in 2015, Cary Fukunaga, who directed the first Netflix original film, *Beasts of No Nation*, told *Business Insider*: 'That's their MO, to not release numbers.'[19] Like any other business, Netflix is entitled to protect its proprietary data. It does now release 'What we Watched' reports, but those barely scratch the surface of what it really knows about our viewing habits.

Going live

Despite its interest in sports team culture and sports-related shows, for a long time Netflix showed little interest in showing live sport. However, other services also traditionally associated with entertainment have got into the game, buying the ability to broadcast the world's most prestigious league competitions, thereby challenging the likes of ESPN and Sky Sports.

Amazon picked up the rights to some English Premier League, Champions League and NFL matches for its Prime Video service. NBC's Peacock shows lots of games from the English Premier League every single week of the season. Most surprisingly, Apple quickly got involved and bought some rights to Major League Baseball and then struck an exclusive global deal with Major League Soccer. The YouTube TV deal for NFL's Sunday Ticket was also huge in the industry.

All of that will be unpacked later in this book. The thing to note at this juncture is that Netflix took years to have any kind of involvement in the live events and sports space. Indeed, co-CEO Sarandos was for a long time rather dismissive of the idea, quipping on a January 2023 investor call: 'We're not anti-sports, we're pro-profits.'

A look at the sums paid by those other sports streamers indicates why it was off-putting to Sarandos and his fellow Netflix execs. Amazon shelled out £30 million per year for the rights to 20 Premier League matches per season, split into two rounds of 10 games for six years. Apple paid out a total of $85 million per year for its two MLB games each Friday night.

Sarandos did later clarify his comments, saying: 'We've not been able to figure out how to deliver profits from renting big league sports in our subscription model. Not to say that won't change. We'll be open to it, but that's where it is today.'[20] Netflix had then seemingly made the decision to not compete for sports rights.

There were various, largely legitimate, reasons for this. For one thing, as the numbers paid by others make clear, acquiring rights to the sports people actually want to watch costs an awful lot of money. Given Netflix has plenty of customers already, why should it expend its capital on sport when it could pay for some more hit series or blockbuster movies instead? Sarandos seemed to be comfortable to let others splash out on the other stuff, while trying to make sure his company offered the best movies and TV shows instead.

On top of the cost of the rights, it is also really hard, technically and presentationally, to get streaming sport right. Fans are very passionate and you don't want to be on the end of a backlash if it goes wrong or they don't like your on-air talent. It was all seemingly more trouble than it was worth for Sarandos and co.

However, just three months after the Netflix boss made his comments about sport, the company did decide to broadcast a live event. In April 2023, it planned to put out a live reunion show of the popular dating reality series *Love is Blind*. It was, to put it mildly, a disaster. Technical faults delayed the live broadcast by over an hour. For many it still did not work at all and they had to watch a recorded version at a later date. Cue plenty of anger online and embarrassment for the streamer.[21] If this had been a live sporting event, with a match going on but fans unable to tune in, the fallout would have been apocalyptic. While Netflix had previously successfully livestreamed a Chris Rock comedy special, the *Love is Blind* fiasco seemed to demonstrate to both the firm and the outside world that some things are better left to others.

But, Netflix did dip its toe back into the water of livestreaming later in 2023. And it did so by doing its first ever bit of live sports broadcasting. It created and showed a pro-am golf tournament in which top professional PGA golfers and enthusiastic F1 racing drivers teamed up and competed for the subtly named Netflix Cup.

The event was tied to two of its own, very popular, shows: *Full Swing*, a documentary series following pro golfers, and *Drive to Survive*, a smash hit show that goes behind the scenes of a Formula 1 season. The Netflix Cup might have been a typically savvy use of intellectual property (IP), but people watched it more for novelty value than high-quality sporting competition. In a similar vein, in March 2024 the company put on the Netflix Slam – a grand title for an exhibition tennis match between Spanish icon Rafa Nadal and his successor Carlos Alcaraz.

These standalone events might not have been serious sport, but it was clear that the tide was turning. Just weeks before the tennis clash, Netflix had made a major announcement. From January 2025, it was going to be the exclusive home of WWE Raw – the wildly popular pro-wrestling sports entertainment show – as well as some associated live events. This marked a major shift from Netflix's previous strategy; although it was a sports entertainment deal, it was not a 'proper' sport (sorry, wrestling fans). WWE is also associated with valuable IP that has a dedicated fan base, so you can see why Netflix might fancy paying up for it. 'Wrestling makes a lot of sense,' says Tom Harrington, Head of Television at the analysis firm Enders Analysis.

But Netflix wasn't done there. The company announced that it would be showing an NFL double-header on Christmas Day 2024 around the globe. This was a shift as dramatic as the introduction of advertising and perhaps underlined the flexibility that has proved to be so important to the company's success. Just because Netflix had previously said it wasn't going to do something doesn't mean it will stay that way forever.

A comment from Netflix Chief Content Officer Bela Bajaria underlined the shift:

> Last year, we decided to take a big bet on live – tapping into massive fandoms across comedy, reality TV, sports, and more. There are no live annual events, sports or otherwise, that compare with the audiences NFL football attracts. We're so excited that the NFL's Christmas Day games will be only on Netflix.

The introduction of both advertising and live broadcasting is connected. Being willing and able to show advertising of some kind makes Netflix open to broadcasting live events and vice versa. The key benefit of buying expensive IP like sports rights is because big brands like to advertise against it as it can encourage high levels of viewership at a specific time.

With the exception of WWE, which is somewhat different, Netflix has largely focused on one-off events, be they novelty pro-am golf tournaments, tennis matches or one day of NFL matches, instead of buying the rights to a whole season of a sport. (As we will see, Apple has taken the opposite approach.)

Some of this is simply down to what was available to buy at different points. Some of it goes back to the company's pre-existing concerns about being in the live sports game and the associated difficulties around negotiating with teams and leagues.

Asked about this in an earnings call in July 2024, Sarandos said the company was making 'Netflix events, not necessarily taking on a lot of tonnage from any one league but actually making these games events, like having two NFL football games on Christmas Day… it really creates a lot of excitement with the service, and it's one day of football.'

Once again, the addition of WWE and NFL was not just about the live action though. Netflix had series related to the sports. It is home to popular shows like *Quarterback* and *Receiver*, which follow major NFL players in key positions, and *Mr. McMahon*, an insightful documentary about WWE founder Vince McMahon. As Sarandos put it: 'We are in love with the kind of very profitable storytelling version of sports. So if you can't wait for those football games on Christmas Day you can watch *Receiver* right now.'

Netflix also played host to a night of boxing that culminated in former world champion Mike Tyson taking on YouTuber Jake Paul. It was a widely marketed event that attracted a huge audience… and also encountered a number of technical issues.[22]

No news is good news

Another area Netflix has been hesitant to get into is news. Some streamers have parent companies associated with news brands. Amazon Prime Video did a live election night broadcast for the 2024 US Presidential Election featuring cable TV stalwarts such as Brian Williams, formerly of NBC, and Shepard Smith, formerly of Fox News and CNBC. Netflix, however, has studiously avoided such a move. Sarandos has repeatedly pushed back against the idea his service would do current affairs, but he also had previously baulked at the idea of doing live sport or having adverts on his platform, and we know how that ended.[23] Such statements are only true until they are not.

And there are indeed hints that the approach to news could change. In October 2024, Chief Content Officer Bajaria revealed that the company was launching a weekly talk/variety show presented by John Mulaney. The popular comedian had previously hosted a week of live shows under the banner 'Everybody's in LA' during the annual Netflix is a Joke comedy festival.

At best, though, such programming is news adjacent. It is highly unlikely that we are going to see a traditional style nightly bulletin on Netflix any time soon, or, probably, ever. Even without this, Netflix, despite being a pioneer in the streaming space, is, by adding adverts, live sport and talk shows, starting to bear an uncanny resemblance to a traditional TV network in a number of ways.

Netflix is a Joke

Stand-up comedy was one of the first things Netflix successfully showed live, but before that the company had long been a supporter of up and coming comedians. In part, this is because they provided (relatively) cheap content to help fill out the platform's catalogue. But its approach to having such shows available, even, perhaps especially, ones that court controversy, has been another key feature of its popularity.[24] Other services offer stand-up comedy specials, but Netflix boasts a catalogue that is better, deeper and wider than those available in many other places – it releases two or three comedy shows each month.[25] Of course, comedians want to sell out the biggest, most iconic theatres around the world. However, in an age when many rise to stardom on social media, having a Netflix special is a huge moment in their career too.

Take Matt Rife, for instance. Around 2022, he erupted on TikTok, largely off the back of sharing clips of his crowd work. Netflix then bought two specials from him. The first, released in 2023, was a fairly standard stand-up comedy set. The other, released in 2024, was him doing the kind of crowd work for which he had become known in the first place. And even though the opening joke of the first special, in which he talks about domestic violence, was highly controversial, the streamer stood by their man and released a follow-up.

Interestingly, a television comedy special might no longer be the pinnacle of a seasoned performer's career in the way it might once have been, but rather a jumping-off point for someone lower down the career ladder. Doing one for Netflix can serve as an accelerant for a comedian's career, propelling them to previously unreached heights. An example of this is Ali Wong.

Wong was a jobbing comedian when her first special, *Baby Cobra*, dropped on Netflix in 2016. The audience kept growing though, even reaching the point where people were reportedly wearing Ali Wong costumes on Halloween! While it's unclear whether or not this is a compliment, it certainly suggests a breakthrough. Most importantly, Wong's future Netflix's specials took place in increasingly large theatres, as she was able to sell out ever bigger venues on the back of the success of the original.

Hannah Gadsby has similarly benefited from appearing on Netflix. Her show *Nanette*, a hilarious and dark exploration of themes such as sexuality, gender and trauma, exploded in 2018, making the comedian a superstar. Maybe talents like Wong and Gadsby would have broken through anyway, but their appearances on Netflix made them known the world over, with word of mouth helping their sets attract millions of views.

Netflix has increasingly played the role of, if not the live open mic night, then a talent scout in comedy. As the examples of Wong and Gadsby show, it aims to highlight the best emerging performers. It does this alongside housing sets from some of the biggest names in the business, such as Dave Chappelle (another act who attracted backlash), Chris Rock, Ellen DeGeneres (post-TV show cancellation and controversy) and Jerry Seinfeld.

While much of this work, particularly with up-and-coming comedians, is not that expensive to produce, becoming dominant in a category involves putting serious money into it. This not only helps attract the best talent, but also deters rivals. One agent told *The Hollywood Reporter* that 'Netflix paid big to get into the space. I don't think the other people wanted to step up to the table.'[26]

Company culture and the keeper test

Despite its ups and downs, Netflix is, by almost any measurement, one of the most successful media and entertainment companies of modern times. Co-founder and chairman Reed Hastings attributes this in no small part to the culture that he and the rest of the leadership created within their firm. It is branded 'no rules rules' and, as you might imagine from such a name, emphasizes flexibility and employee responsibility.

So central to the firm is this culture that there is a deck of 125 slides that explicitly lays it all out. A version of it was first published online in August 2009 and subsequent updates have also been made public on the Netflix website. Corporate jargon like 'honest, productive feedback', 'dream team' and 'freedom and responsibility' appear throughout the document. It starts with five key bullet points that are worth reading in full:

1 Encourage decision-making by employees

2 Share information openly, broadly and deliberately

3 Communicate candidly and directly

4 Keep only our highly effective people

5 Avoid rules

You may well ask why it requires 125 slides to tell people to avoid rules, but that rather sums up what 'no rules rules' is all about.

Elsewhere, the document explains the 'keeper test', something a number of former Netflix employees have highlighted adds both to the calibre of staff at the company and the constant stress they are under.

The keeper test is something managers are expected to consider constantly and is defined as follows: 'If a team member was leaving for a similar role at another company, would the manager try to keep them?' The document goes to explain that 'those who do not pass the keeper test (i.e. their manager would not fight to keep them) are given a generous severance package so we can find someone even better for that position – making an even better dream team.'

This could all easily be dismissed as tech bro speak or more corporatese, and perhaps it should be. It is the kind of thing that immediately puts unreasonable pressure on even very good employees. Furthermore, it might potentially put off women as well as those from diverse communities and different backgrounds applying to work at Netflix, who may have additional responsibilities or barriers to entry but could still perform at an extremely high level and bring benefits to the company. Netflix's success does, though, indicate that the culture and methods of achieving it also have plenty of benefits, not least by encouraging a constant desire to be better and by pushing staff to take risks and innovate.

Can the rest of us learn something from it? Hastings certainly thinks so. In the introduction to a book on the subject that he co-authored with academic Erin Meyer, he writes: 'Our culture, which focused on achieving top performance with talent density and leading employees with context not control, has allowed us to continually grow and change the world, and our member needs, have likewise morphed around us.'[27]

In reality, Netflix thinks of itself as offering the antithesis of the Silicon Valley culture. While the company insists it offers plenty of perks for working there, discussions and documents frequently reiterate that the workforce is not a 'family', an approach that is prevalent in startup land. Instead, Netflix sees itself more like a sports team, something the 'keeper test' reinforces. It is the equivalent of a Premier League manager asking if they have the best centre forward they can possibly get or whether they can bring in a more prolific goal scorer in the next transfer window. As with elite sports, this leads to a highly stressful work environment, with staff at all levels frequently looking over their (well-remunerated) shoulders.

Over the years, Hastings has even let go of some of the senior executives closest to him as they eventually came to fail the 'keeper test'. He also insisted that the board could apply the principle to him, although there was no indication that he had failed it when, in January 2023, he stood down as CEO and became chairman.

Game on

Having made a huge impact in the world of television and movies, Netflix next set its sights on gaming, launching Netflix Games in November 2021. The games are accessible to subscribers via their mobile devices and many relate to shows and films available on the service.

While it was clearly a bid to once again innovate and stay ahead of the competition, the initial signs were not that positive. The games had not proved to be all that game-changing. In August 2022, data from analytics company Apptopia revealed that they had been downloaded 23.3 million times and were averaging 1.7 million daily users. At that point, Netflix had 221 million subscribers, meaning that less than 1 per cent of the most likely players were actively using the Netflix Games offering per day.[28]

The user experience is not a particularly good one, which probably doesn't help. While Netflix makes it exceptionally easy to find things to watch, playing games is a bit clunkier. They are listed within the app, tap and you can find something you want to try. But users are then directed back to their phone's usual app store. It is not abundantly clear why I need a Netflix subscription in order to play solitaire on my phone.

On the other hand, having games linked to Netflix IP does make sense. This seems to be the particular focus of the company and executives have spoken publicly about it being part of a superfandom offering for those particularly devoted to certain series.[29]

Having insisted that it was experimenting and moving relatively slowly in the gaming space, the company has seemed to get less interested in this offering over time. While it persists with new releases, games were mentioned just once in the October 2023 letter to shareholders, compared to the nine times it had come up in the same quarter the year before. At that point Hastings insisted that 'we're seeing some encouraging signs of gameplay leading to higher retention'.[30] This seems, at best, optimistic.

Disruption

It's hard to overstate the level of disruption Netflix has caused to the entertainment industry – for good and bad. Its originals rack up hours upon hours of viewing time each week. According to its Most Watched data released at the time, by October 2024, its most popular English language movie, *Red Notice*, starring The Rock, Gal Gadot and Ryan Reynolds, had clocked over 230 million views over more than 454 million hours of viewing time. Gothy drama series *Wednesday*, *Stranger Things Series 4* and *Dahmer Monster: The Jeffrey Dahmer Story* had all accrued a billion hours of viewing time.

Netflix put jobs into Hollywood, allowing people who may never had got a chance to put their work out in front of a global audience. The company was, certainly in its early days, prepared to take risks on the kind of programming it commissioned and it continues to provide a platform for films and TV shows from all over the world, some of which would surely never have been made without Netflix's resources to back them.

There is no doubt that Netflix also puts huge pressure on writers and actors and that such practices were a major factor in the Hollywood strikes of 2023. By introducing adverts it rowed back on one of its key promises to its audience.

Whatever your view of the company, it is indisputable that Netflix changed TV viewing habits, prompting plenty of rivals to try and copy what it was doing and steal its streaming crown. Some of those companies weren't traditionally even in the media business…

02
The arrival of rivals

How tech became media and media became tech wars

Hollywood is littered with august filmmakers, titans of industry and companies that have been around for a century or more – the likes of Disney, Warner Brothers and Paramount. These firms and their executives have long been the power players in Tinseltown, helping to set cultural trends around the world.

Streaming has blown up TV and film to perhaps an even greater extent than digital revolutions have impacted other sectors. As it became ever more popular, Netflix got people around the world used to this on-demand way of consuming media. However, it was still relatively simple for the old guard to maintain their position when there was only one big streamer to deal with. Then Big Tech decided to get involved. Eventually, those once untouchable Hollywood firms realized they needed to adapt. A number arguably did so far too late.

The likes of Amazon and Apple are some of the most valuable companies on the planet. Their bank accounts allow them to do pretty much whatever they want, in whatever sector they want. That doesn't mean products and projects don't eventually have to make a profit, but they can afford to take a lot of swings before having to walk away from the plate. The traditional studios are simply not in the same position. By getting into the entertainment business, these tech firms became cultural phenomena, not just the makers of widgets. They expanded their brands, embedding themselves even more deeply into all aspects of our daily lives.

There is a strong argument that having a range of streaming platforms available, whether they be from tech forms or the more traditional players, has been good for viewers. They have to compete in terms of both content and price. We certainly have plenty to watch now! However, the viewing environment has become very fractured and consumers are asked to make more and more monthly payments in order to access everything they actually want. This can certainly be frustrating. People regularly complain that they don't know where to find what they want to watch and/or cannot be bothered to pay for another service. Yet that has not deterred a whole host of companies from launching their streaming products. We'll look at some of them in this chapter.

Amazon

Amazon started selling books online in 1995 and then moved into selling… everything… [1] Eventually, this included offering a streaming service. The company had actually made video content available as far back as 2006, but it was the arrival of Prime Video that thrust Amazon into the streaming wars.

Bundled into its Prime subscription service, Amazon allows customers both free one-day delivery and a host of content for a pretty reasonable (albeit increasing) price. For many, the shows and movies were, and still remain, little more than an add-on to the brown cardboard boxes that turn up on your doorstep just hours after being ordered. However, the sheer size and scope of the company means that it is now a significant combatant in the streaming wars.

Launched in 2011, it was boss Jeff Bezos who reportedly came up with the idea of making the streaming service part of a package designed for his company's most committed customers. Some of the executives involved have even admitted that at the beginning the content on Prime Video wasn't particularly good.[2] That didn't really matter at that point as people felt that they were getting some content essentially for nothing. When looked at in that light, whatever they were watching was perfectly sufficient. Now, of course, quality does very much matter. Amazon is now a multidimensional behemoth that makes hit TV shows and regularly broadcasts popular sporting events.

Tom Harrington, Head of TV at analysis firm Enders Analysis, believes that Prime 'is the stickiest thing ever' – i.e. once people are in they are likely to stay, largely because of the free shipping. He does not think the connection between shopping and streaming is all that strong.

'There seems to be no connection between video and the e-commerce part' of Amazon, he says.

The free, fast shipping is undoubtedly a huge draw that still encourages many to take out a Prime subscription. However, some at Amazon claim there are plenty of instances when the reverse is true too. They argue that shows and movies can work as a hook to get customers into the ecosystem and the convenience and benefits of the overall offer keeps them there – the stickiness Harrison speaks of.

'What we find,' Amazon executive Jeff Wilde once explained in an interview, 'is that customers who watch a movie that they love, they buy more Tide [laundry detergent]. They shop more, they're more likely to renew their Prime subscription, they're more likely to convert a free trial into a monthly or annual Prime subscription.' He concluded that 'ultimately, viewers are telling us with their actions that video is an important part of the Prime experience.'[3] It is the combination of the shopping capacity and the content that make Prime such a powerful weapon for Amazon in the streaming wars.

Purchasing Bond, James Bond

One very symbolic moment in the development of Prime Video was when Amazon purchased MGM, the film brand that is home to iconic franchises, most notably James Bond. The $8.5 billion acquisition was first announced in May 2021 but regulatory rows meant it took another 10 months to close. A tech giant buying a Hollywood studio that had been around since 1924 perfectly demonstrated the upheaval the entertainment industry was undergoing. Mike Hopkins, Senior Vice President of Prime Video and Amazon Studios, made no bones about why his company was interested in the deal. Commenting when it was first announced, he said: 'The real financial value behind this deal is the treasure trove of IP in the deep catalogue that we plan to reimagine and develop together with MGM's talented team. It's very exciting and provides so many opportunities for high-quality storytelling.'[4]

The European Commission, it should be noted, was somewhat less convinced about the significance of this deal. Giving its unconditional approval in March 2022 following an investigation, it said: 'Even in the national markets where Amazon has a sizeable market presence among video streaming platforms, the Commission found that Amazon faces strong competition from other players.'

The Commission added that they 'found that MGM's films represent only a limited share of box office revenues in the EEA [European Economic Area] and that overall MGM is not among the top production studios, despite its rights over successful film franchises such as James Bond'.[5]

Whatever EU officials might think, with its purchase of MGM, Amazon had picked up the rights to all sorts of popular movies and shows without having to make them itself. Prime Video has gone on to position itself as the Home of Bond, making it easy for fans to watch the entire catalogue. It is also offering additional 007 content. If you're a James Bond fan, Prime Video is the place for you.

Of course, this is not the only reason Amazon splashed billions of dollars on the deal. Classic movies such as *Robocop*, *Rocky*, *Silence of the Lambs* and *Thelma &*

Louise, as well as series such as *Stargate*, also came as part of the package. Whether it is down to people rewatching the original work or by making sequels and creating merchandise to be sold via Amazon, the company can and will make money from this content for years to come. The company's financial might allowed it to bypass many of the stages traditional studios had to go through for decades. It didn't need to create and market original IP, instead it just bought some of the best-known movies on the planet. Owning this content gives the streamer an air of credibility too, something that it perhaps needs given we still mostly think of Amazon as a shopping website.

Not very original

Buying already popular work was appealing to Amazon became it has been less successful than its rivals in creating hit original shows that become huge hits. In 2017, it spent $1 billion creating a series based on the *Lord of the Rings* books, already the basis of a wildly successful set of films which the company thought it could cash in on. When this series was announced, some noted that Amazon had not had a hit to compete with the likes of *Stranger Things* on Netflix.

Reacher was among the Prime Video series that proved to be popular, centring on a character, Jack Reacher, who had been created in novels by Lee Child. The books were the basis of the series. Another all-action Jack, this time Jack Ryan, also featured in a series that did well on Prime Video, and that was also based on books, this time by Tom Clancy.

All streamers lean heavily on work based on characters and stories created by authors, and authors are, quite rightly, happy to be paid for their creations being used in this way. It is a particularly useful way to do things if you're a company associated with selling books, but it is not exactly pushing any creative boundaries. (Apple, which has its Apple Books platform, also makes a significant number of adaptations.) Amazon has then made shows people like to watch, but perhaps not content that has captured the zeitgeist in the way output from some of its competitors has.

To a certain extent, the same is true of the other companies discussed in this chapter. This is because streaming is not and almost certainly never will be a core part of their business. If they make content that customers want to watch, then that is an added bonus.

Moving into live sport

The originals might not be shifting the dial too much but, as we will explore further in the next chapter, one area in which Amazon is really making an impact is live sports. In the UK, this included buying the rights to show some Premier League football. Amazon then added some Champions League football matches to its roster. In the USA, it picked up the rights to Thursday Night NFL matches. Prime Video is also a hub for tennis coverage in multiple countries. These are all meaningful assets that attract people into taking out subscriptions of various kinds. If you want to watch this content (legally), you have no other choice but to pay up and make your own small contribution to Jeff Bezos's 'going to space' fund.

And if that wasn't enough, subscribers can access other kinds of live television via Prime Video. Much of this is low-grade channels dedicated to repeats of shows, but it does include some high-quality news content such as CNN and Bloomberg. You can also take out subscriptions to various other services within the Prime Video app including Premier Sports and MGM+.

While series like *Reacher*, and indeed James Bond movies and NFL matches, are good television, and Prime Video is a useful platform that demonstrates Amazon is taking being in the media space seriously, we should not be fooled. Selling and delivering stuff is still Amazon's fundamental business. This sits alongside Amazon Web Services (AWS), which is a profit-generating machine all by itself.

Those involved at the top echelons of the company would undoubtedly dispute this in public, but the reality is that primarily Amazon wants to be the place where you do all your shopping and build the cloud computing infrastructure for your business. If people can get some good entertainment from them as well, then great. However, Amazon is and always will be primarily a logistics and technology company. Likewise, Apple is primarily a company that makes and sells phones and computers.

Apple

'A billion pockets y'all,' shouts Oprah Winfrey.[6] She's standing in the sun with the likes of Jennifer Aniston and Reece Witherspoon. It's 25 March 2019 and earlier in the day Apple had finally revealed details of its much-anticipated streaming service. It did so with a star-studded event that included such Hollywood luminaries as Oprah, Steven Spielberg and Sesame Street's Big Bird. All the streamers like to use star power, but it was certainly an impressive cast list that hinted at what Apple might be capable of pulling off. In her 'billion pockets' comment, Oprah was referring to the number of iPhones that had been sold, rightly noting how important that was in getting people to watch the content. Apple TV+ went live in November 2019.

The assembled celebrities aside, what marked out Apple's event is that it revealed that its service would, apart from a few classic *Peanuts* specials, contain only original content. As we've seen, other services took a long time to start making their own series and films but Cupertino-based Apple had chosen not to get involved in catalogue battles and instead splash its cash on brand-new work.

Even before it rolled out the red carpet, we had an indication of how seriously Apple was taking its jump into the entertainment industry when Jamie Erlicht and Zack Van Amburg joined the company. They had led Sony Pictures Television since 2005 and have been behind popular shows like *Better Call Saul, Breaking Bad* and *The Crown*.[7] At this point, we still did not really have any idea of what the company's streaming offering would look like. However, it was clear that these two, and by extension their employer, meant business. Erlicht and Van Amburg had led Sony Pictures Television since 2005 and had been behind popular shows such as *Better Call Saul*, its predecessor *Breaking Bad* and *The Crown*.

Apple TV+ was initially met with scepticism. People wanted to know they why they needed to sign up for yet another streaming service, especially one with no known content in its catalogue. Paying to be able to watch *Friends* on demand is one thing, but nobody had ever heard of the shows on TV+.

This is where Apple was able to deploy the full might that comes with being, depending on the day, the most valuable company in the world. It declared that anyone who had recently bought or went on to buy new Apple products would get a year of TV+ free. In reality, this meant that almost nobody paid for TV+ for at least a year, as most of those who wanted it fell into those categories. The offer gave people a chance to see if there was anything on the platform worth watching.

Enders Analysis's Tom Harrington remarks that Apple's approach to streaming is seemingly 'We're gonna have a big show every six months and no one's paying for it.' He wonders what the company's overall strategy is. 'Is it to sell more iPhones? Is it just to bring people into a video marketplace that can sell subscriptions to other products?'

Apple was giving away the expensive content. This would be impossible for traditional media competitors, for whom entertainment is their key asset, to do. It is similar to what happened at Amazon, which was able to bundle its content in with its pre-existing free delivery service. Ultimately, the tech giants could afford the talent and could also afford to make it cheap for people to watch – a powerful combination of advantages to have when launching a service.

Apple also offers a services bundle. Called Apple One, it is the equivalent to Amazon Prime. There are different tiers at different price points available, depending on which products you want access to, but for a customer who wishes to use a number of them, buying an Apple One package is the most cost-effective way to do so. Crucially, TV+ is available at every tier, so even those who might really just care about having iCloud storage and music will also get TV+ and the chance to sample its shows. This is yet another demonstration of the advantages the tech firms have when entering into the streaming wars. They have an array of services to offer and the financial capacity to bring them together at a discounted price.

Who is watching anything on Apple TV+ (apart from Ted Lasso*)?*

Writing this a few years into the experiment, it's hard to know if Apple's strategy is paying off. Tim Cook and Apple executives are not exactly forthcoming with numbers and there are regularly free trial offers available, making it easy to watch popular shows at no cost. It is pretty clear though that Apple TV+ is by no means the largest streaming service, not least because it hosts considerably less content than its rivals. 'If it's just entirely a brand-new thing, and it's just how many Emmys we've won, then it's been a success,' says Harrington. It has won prestigious awards but the reality is that if Apple TV+ had a huge subscriber base the company would be shouting about it.

However, many people are actually watching, the service does feature some of the world's biggest stars and has also had some enormous hits, most notably *Ted Lasso*. Based on an advert by NBC Sports originally released in 2013, the series tells the story of a hapless American football coach (Lasso, played by Jason Sudeikis,) who ends up in the UK coaching a professional football (soccer) team. Released in the depths of the 2020 Covid-19 pandemic, the feel-good show was seemingly exactly what everyone needed, as Ted spread his wit, charm and wisdom to win over his players and their fans. While high-budget, well-trailed shows like *The Morning Show* and *For All Mankind* had launched the service, it was Ted Lasso and his friends and colleagues at AFC Richmond that made many people pay attention to Apple TV+. The series remained immensely popular throughout its initial three-season run and remained near the top of the Apple TV+ Top Chart long after it finished. Such was the demand, *Ted Lasso* came back for a fourth season. This is despite the fact that creator Bill Lawrence told me in an interview I did for *The Mac Observer* that it had been written as a three-season arc.[8]

Winning big

In typical Apple fashion, the firm has proved itself to be a late but influential entrant to the streaming space. Notably, it was the first streamer to win a Best Picture Oscar, claiming the prize for the movie *Coda* in 2022. In doing so, Apple TV+ achieved something that Netflix had been trying to do for decades, albeit with a film that it purchased at the Cannes Film Festival instead of one made in-house. Troy Kotsur was named Best Supporting Actor for the film at the same ceremony, becoming the first deaf actor to pick up the accolade.

Despite the fact it was a purchase from a film festival, the Oscar success for was still a huge moment. It put a tech company right at the heart of the entertainment industry, allowing Tim Cook and his colleagues to put on smart suits and go to swanky events. As Erlicht noted in a statement after the 2022 Academy Awards: 'What an incredible journey it has been since the moment we first saw *Coda* to today's historic recognition from the Academy.'

Away from the big screen, Apple, like Amazon, has decided to purchase the rights for TV+ to show live sport. Again, the company did this much earlier than other streaming services did. The first move was when it started to show Major League Baseball on a Friday night. The deal with MLB was announced in March 2022 and came as something of a surprise, not least because Apple is a notoriously cautious company and broadcasting live sport is an expensive risk to take. It then went on to announce a 10-year deal with Major League Soccer. In July 2025, it emerged that that Apple was set to win the rights to show Formula 1 in the US too. (More on all this later.)

YouTube

Arguably, the most successful streaming service of all is YouTube. It might not be home to blockbuster movies or epic TV series in the way Netflix and the like are, although you can buy and rent such work via the platform, but there are a huge number of traditional broadcasters that would kill for YouTube's viewing figures. Millions of hours of content are uploaded to the platform each day, with the biggest creators able to attract hundreds of millions of views. Most incredibly of all, YouTube barely pays for or commissions any of this work. People willingly hand it over in the hope of achieving stardom and the income associated with it.

YouTube offers a revenue sharing model whereby creators who meet certain criteria – 500 subscribers, 3 video uploads per 90 days and either 3,000 viewer hours in a year or 3.5 million shorts views in 90 days – can monetize their account. The Google-owned platform offers a 50/50 split on revenue generated by long-form content and a 55/45 split on Shorts content. This model, called the YouTube Partner Program, is far more generous than those offered by the likes of Instagram and TikTok and has been around since 2007.[9] Getting approved to the programme is always the first ambition of wannabe creators, even before they start trying to tie up brand deals – sponsorships from which they can also earn.

On the back of the Partner Program and other monetization tools, a number of creators have built success on YouTube, making a decent living through the platform. Some have even earned enough to build studios and staff similar to that of TV channels. No wonder that survey after survey has flagged how young people desire to become influencers and creators instead of taking up traditional jobs. A 2023 Morning Consult survey of 1,000 Gen Zers (those born approximately between 1996 and

2010) found that 57 per cent wanted to be influencers. Furthermore, 41 per cent of all adults said the same thing.[10]

If streaming services broke the mould of linear TV, YouTube broke the mould around media and publishing almost entirely. It has its own language, culture and norms. From homemade vlogs to multi-camera productions, YouTube allows everyone to publish their creative ideas without having to go through any gatekeepers. As author Chris Stokel-Walker put it in *YouTubers*, his seminal book about the platform: 'YouTube is a kaleidoscope of visual and audio content that mimics the richness, quirkiness, beauty and madness of human life.'[11]

There is no need to haggle with agents, pitch a script and get taken on by a studio. If you have an idea you can make it and put it out into the world to see if people, and the YouTube algorithm, like it. Ultimately, it is a staggering achievement for any kind of online platform to have lasted for multiple decades. Not only has YouTube achieved that feat, but it has also become more, not less, relevant as time has gone on.

Beastly problems

Perhaps the most popular YouTuber is Jimmy Donaldson, better known around the globe as MrBeast. His highly produced, outrageously elaborate output attracts a staggering viewership. A 25-minute-long video called '$456,000 Squid Game In Real Life!' got 782 million views in the three and a half years following its release in November 2021.

Building on this, Donaldson also secured a deal to make a show for Prime Video based on the over-the-top challenges he and others undertake in his YouTube videos. It is a fascinating example of a star from one platform collaborating with another, in this case in a deal thought to be worth $100 million.[12]

However, things were not without problems. A *New York Times* investigation revealed that contestants in the selection process, which was filmed for the MrBeast YouTube channel, involved 2,000 people. That was twice the number the contestants thought would be taking part when they applied, although the YouTuber's team insisted this was always the plan. Worse still, contestants reported having barely any food or sleep and struggling to access the prescription medications like insulin that they required. Female contestants who were menstruating also recalled staff not taking their requests to access some of their clean underwear seriously.[13] The incident is, perhaps, a stark reminder of how difficult the jump from bro-tastic YouTuber to serious TV production is.

The new TV

The work put out by Donaldson and those who aspire to be him is a crucial element of the streaming wars. It vies for viewers' time and attention, as well as advertisers' cash. In fact, for many, YouTube is now TV. Engagement with the platform increased by 27 per cent in 2024, according to research by MoffettNathanson, while the top tier of streamers combined only went up 8 per cent.[14] A huge part of this growth is surely due to the platform becoming available on smart TVs. It is no longer about random videos you find on your computer – it is there alongside Netflix, Prime Video and whatever cable package you pay for.

As well as young upstarts, the power of YouTube has helped it attract major names from mainstream media. Megyn Kelly was a hugely popular talk show host on Fox News in America, even, infamously, hosting a Republican presidential debate in which she questioned Donald Trump on his comments about women. After this incident and revealing sexual misconduct by former Fox News boss Roger Ailes, Kelly left the channel for a daytime career on NBC that ended when she questioned whether blackface was wrong.[15] Far from being the end of her time broadcasting, she has gone on to set up an independent media company and have a popular podcast via SiriusXM, which also appears on YouTube. That YouTube channel has millions of subscribers.

Piers Morgan, meanwhile, walked away from being the highest-paid journalist in the UK to build an empire around YouTube. His *Uncensored* show had moved exclusively to the platform following the 2024 collapse of TalkTV in the UK. In 2025, Morgan decided to buy the channel outright instead of being, as he put it in one interview for *The i Paper*, 'talent for hire', adding 'I wanted ownership'.[16] Since the move to independence, Morgan's channel continued to put out fiery debates attracting hundreds of thousands of views, sometimes more.

There are also the likes of Don Lemon and Mehdi Hasan. Both lost jobs at big cable networks and went to YouTube to rebuild. (Hasan also used Substack to build his Zeteo brand.) It is clear that after years of the restrictions imposed by traditional broadcasters, all of these presenters like the freedom offered by digital platforms, and YouTube in particular.

YouTube is not just about content from creators, podcasters and brands either. In 2017, it launched YouTube TV in the US. The service allows users to stream major networks like ABC, CBS, MSNBC, Fox News, Fox Sports and many others at a pretty competitive price ($83 per month in 2025). It has added things like the NFL's Sunday Ticket set of games too for an extra cost, making it an even more attractive proposition. Now, whether you want to watch the latest clips from beauty-influencer Zoella, a monologue from Megyn Kelly or breaking news on CNN, you do not have to leave the YouTube app.

The first video ever published on YouTube featured co-founder Jawed Karim and was called 'Me at the zoo' and, as the title suggests, it was a grainy clip shot at San Diego Zoo. YouTube has thrived in the decades since its launch. Its longevity is not just down to being part of the Google behemoth, it is about its ability to adapt and support creators through attractive revenue-sharing programmes and by building new functionality, such as YouTube TV. As well as being a frontrunner in digital video, it typifies the era in which media and tech have collided and become one.

Tech becomes media

These moves by tech firms are not just about people watching their content on the services that they run. Both

Amazon and Apple also want to make their apps and hardware the hub that you use to consume *all*, or at least most, of your content. They provide a platform to take out subscriptions to other services which you can then access through their own app and/or hardware. You can subscribe to Paramount+ via both Amazon and Apple, for instance. A classic name in the film and entertainment industry, Paramount has chosen to play nicely with the tech titans, allowing subscriptions to its streaming service via Apple's App Store and Amazon's Prime Video channels.

These arrangements demonstrate just how important the tech infrastructure is to streaming and what a stong position the likes of Apple and Amazon are in. The deals with others not only keep users in their ecosystems, but it means they take a cut too. As they already have your payment information, it is a very simple process that makes you more likely to make the purchase and forget to cancel a subscription. It is arguably a win for all concerned, although one always has to be wary of any platform becoming too dominant.

Rivals come together

In a quite unexpected move in 2024, Apple made TV+ available to access through Prime Video's Channels facility in the US and then in the UK.[17,18] The move was clearly Apple investigating the concept of making its original content available on rival platforms.

In some ways, the move seemed like something of a surrender, an admission by Apple that it simply was not getting enough eyeballs in its own app. Why else would

Apple, a notoriously territorial and defensive company, not force people to watch its content directly on its platform? The move certainly seemed in contradiction to Apple's apparent desire to turn its TV app (as opposed to the similarly named service or the identically named hardware box) into a streaming hub.

With Tim Cook as CEO, the company is nothing if not pragmatic. The leadership had clearly decided that they needed to boost uptake of the service by making TV+ more readily available, even if it meant collaborating with a bitter rival. Notably, at the time of writing this book, the arrangement was not reciprocal and Prime Video remained unavailable via Apple's TV app.

Disney magic

The traditional media company that put up the best fight against the insurgents is Disney. This is somewhat unsurprising. As famous as Warner Brothers and others are, there is arguably no more prestigious or well known a media brand than Disney. Despite this, it took its time to launch a streamer, leaving its content available across all manner of services for a number of years.

When it finally launched Disney+ in 2019, just days after Apple had launched TV+, Disney was entering the streaming market from a position of immense strength. It had a vast catalogue of movies and shows spanning, at that point, almost a hundred years. It underlined this by making the first ever Disney movie, *Steamboat Willy* starring Mickey Mouse from 1928, available on the platform.

Then we have the modern Pixar movies. Who wouldn't sign up so that their child can watch *Moana* for the millionth time? In addition to all that, its ownership of astronomically popular brands like Marvel and Star Wars meant that Disney was the closest to having guaranteed success with its streaming service at launch.

Disney is also a multi-revenue streaming business. It has theme parks, stores and so much else. This means that it could launch its streamer in a similar way to the tech companies. Although Disney+ is undoubtedly far more important to its parent company than its equivalents at Apple and Amazon, it did not immediately need this new service to rescue the whole business. Of course, the streaming service had to work in the medium to long term, but other elements of the company, such as the theme parks, allowed Disney+ a little bit of breathing space.

Because Disney+ launched at a similar time and with a similar price point to Apple TV+, it allowed for some interesting comparisons. Disney's fiscal year 2024 results showed that by November 2024 it had over 120 million Disney+ Core paid subscribers. It's almost impossible to know what that number is for Apple TV+, but estimates put it at around 25 million, with a lot more subscribers using the service for free.

Disney called on a rival for tech support as it grew its streamer. Disney+ is built using AWS servers from Amazon. In the streaming wars, everyone is a tech company and a media company. Everyone is both an enemy and ally.

Disney has long had a relationship with Apple too. Its CEO Bob Iger was a close confidant of the late Apple founder Steve Jobs and only left its board on 10 September 2019 when Apple TV+ was about to launch. The conflict of interests was obvious and insurmountable.[19] As Iger put in on CNBC: 'our paths were conflicting rather than converging'.[20]

The relationship remained though. Iger even appeared at the event at which Apple launched its Vision Pro headset, with Disney+ being one of the apps built into it. It also created a unique Vision Pro experience for season 3 of its animated series *What If…?* Netflix has shown little interest in doing similar.

Strategic alliance

Another significant deal involving Disney+ was announced in July 2025. Free-to-air British broadcaster ITV revealed that it had teamed up with the Walt Disney Company to make a selection of each other's content available on both platforms. Going live on 16 July 2025, it was branded A Taste of Disney+ or A Taste of ITV, depending on where you are watching.

This was about more than just sharing the unwanted leftovers. Huge ITV hits like *Mr Bates vs The Post Office*, which finally sparked public and political attention on the Post Office scandal, and thriller *Spy Among Friends* were part of the deal. So too were some seasons of *Love Island*, the reality show that helped sell thousands of water bottles.

On Disney's part, seasons of multi-award-winning drama *The Bear* were included in the arrangement, as were parts of the *Star Wars* universe such as seasons of *Andor*. Reality hits like *The Kardashians* and *The Secret Lives of Mormon Wives* also became available to ITV viewers.

The companies hoped that making their work a bit more widely available would prompt viewers to tune in more (in the case of ITVX, which can be accessed for free) or buy a subscription (in the case of Disney+). That's why only some seasons of certain shows were available cross platform. Joe Earley, President, Direct-to-Consumer, Disney Entertainment, made that clear in his quote in the press release. He said that the collaboration would 'encourage ITVX viewers to discover some of Disney+'s award-winning series and blockbuster films'.

Meanwhile, in the same release, Kevin Lygo, Managing Director of Media and Entertainment at ITV, described it as a 'mutually beneficial alliance' and said it 'allows us to show our complementary audiences a specially selected collection of titles, regularly updating, that gives a flavour of the range in our respective offerings'. He added the result would be 'even more great content for viewers on ITVX, and even more opportunity for viewers to find and enjoy our distinctive titles and services.'[21]

The arrangement was a strategic partnership described by the two firms as the 'first of its kind'. For example Sky customers could already receive some Peacock TV programming, but Sky and Peacock have the same parent company (Comcast). The ITV–Disney+ deal is a significant arrangement between two totally separate media brands.

It hinted that rebundling – when a group of services come together in a package that includes them all at one monthly price – might be speeding up. This has not become widespread at the time of writing, but streamers were starting to realize that customers were not going to take out infinite subscriptions. Collaboration with others was needed to get eyeballs in front of your work, as well as advertising and at least some subscriber money behind it.

That's not the only reason this deal, although extremely significant, was not necessarily surprising. The two companies already had an arrangement through which ITV made Disney+ shows like *Renegade Nell* available free-to-air in the UK. ITV studios also served as the producer of Jilly Cooper adaptation *Rivals*, as well as *Suspect: The Shooting of Jean Charles de Menezes* and the long-awaited, albeit arguably not entirely necessary, return on Disney+ of dating show *Blind Date*.

The announcement was also unlikely to be a one-off. Karl Holmes, Disney+'s General Manager for Europe, the Middle East and Africa, told industry publication *Deadline* that he expects 'we will announce similar deals with other broadcasters [across EMEA] in the coming months'. He added that 'there will be lots of different models that achieve a similar outcome' – i.e. spreading content more widely.[22]

The streamers might continue to battle against each other, but strategic alliances are becoming more and more important.

Speeding up FAST

Samsung, the leading TV maker in the world, has a service called Samsung TV+. (You may be noticing a theme in the services' names at this point...) It is not a streaming service in the Netflix sense. You can't pick a specific episode of a specific show and start watching it. Instead it is what is called a FAST service – free ad-supported television. It has a list of channels that you can scroll through and these are paid for via adverts instead of subscription. This concept might just sound familiar – it resembles the linear TV setup we all grew up with.

There are other services, such as Pluto, which are free and combine live and on-demand offerings alongside adverts too. While the growth of FAST services is fascinating, the important thing to emphasize at this juncture is that Samsung TV+ is curated and made available by Samsung, the tech company. The very same firm that makes and sells the hardware that millions upon millions of people around the world have in their homes has a platform through which you can consume content that it has selected. It's not just on TVs either. The service is accessible via the company's mobile devices and one of its range of fridges too. Really.

Samsung TV+ actually landed in 2015 as a rental service before taking on its current structure. It arrived pre-installed on Samsung TVs from 2016.[23] We are not talking about high-end TV here. There are various channels dedicated to things like repeats of Graham Norton's and Conan O'Brien's talks shows, sitcom series and quiz shows. However, as with Prime TV channels, there is some content that many would pay for, notably news through the likes of CNN, CNBC and Bloomberg TV+. There is even occasionally live sport on some of the available channels.

While I doubt the existence of Samsung TV+ is influencing many purchasing decisions, it's some extra stuff you get free with your phone, tablet, TV or… fridge, making the device just that little bit more valuable. Interestingly, rival hardware makes have not made a similar move to the Japanese giant.

Owning the ecosystem

Control of the infrastructure and ecosystem through which we stream content is a huge weapon in the arsenal of the tech companies. It is also one of the areas in which Netflix and the traditional media companies cannot and will not challenge them. Netflix is not going to make laptops or phones. It is unlikely to make a TV or streaming stick like the Amazon Fire products either, nor does it appear interested in being a platform through which you watch other people's work. The whole point is that its catalogue makes a subscription valuable, not deals with others.

At its core, Netflix is a media company. Yes, it absolutely has to focus on the technology on which its service operates. Its app and related infrastructure are crucial to its success. However, it isn't Apple or Amazon, nor does it want to be.

Media becomes tech

As the streaming wars have escalated, the traditional media companies have had to adapt and take on at least some of the characteristics of tech firms. Building a direct-to-consumer (DTC) offering requires serious infrastructure. It also requires a change in mindset, away from the big blockbusters and one-off payments towards having a catalogue that people want to consistently pay for.

Arguably, many of these companies spent too long holding on to the old models, hoping people would keep buying cable and return to the cinema. Consequently, they have had to play catch-up having finally realized they need to change tack. By that point, people were already querying whether they needed yet another streaming service.

Furthermore, launching such a streaming service was far from a guaranteed win for many of the old guard. It was a competitive market and, as we have seen, they are fighting against powerful rivals. You have to work hard to squeeze as much revenue as possible out of every customer. This is something one media executive in particular has battled with.

Warner Bros tries to discover streaming

In 2022, two major media companies – Warner Bros and Discovery – came together to form a new entity following a protracted and much analysed merger process. When he took over as CEO of the newly created Warner Bros

Discovery, David Zaslav was faced with a number of issues, not least how to make money from streaming.

Zaslav himself is a fascinating figure. He is, to some extent, an unlikely modern media mogul, a man arguably better at number crunching then creativity. Nevertheless, when he got his hands on the new role his focus was to try and drag the company into the streaming future and out of the red. He undertook a major cost-cutting exercise, which included ending the TV series *Westworld* and cancelling long-awaited movies such as *Batgirl*[24,25]. It was cheaper to write off the cost than release them, however disappointed fans might be. This was far from the end of the story though.

In June 2025, Warner Bros Discovery announced that it was once again becoming two separate companies. Significantly, one of these would be a streaming and services firm, while the other was to encompass all of the linear TV networks and their associated products. David Zaslav was assigned to lead the former, making it clear where the priorities were going to lie.

Streaming service HBO Max was a central pillar of the company set to fall under Zaslav's leadership. All in all, the merger from three years before was essentially undone by the new arrangements, although connections remained. The linear TV company, initially named Global Networks, was to have a 20 per cent stake in its sister streaming entity, with the expectation being that this would largely be used to pay off debt, a problem that hounded Zaslav and WBD.[26]

Building out profitable and effective streaming services was central to the plan. This doesn't just apply to the streaming and services company itself. CNN has struggled to build a presence in streaming, including the aborted and frankly embarrassing attempt to launch CNN+, which closed down within a month in 2022.[27] The announcement of the split made it clear that the news giant had to find a streaming solution. (It's part of the reason why Mark Thompson, a much-lauded executive previously at the BBC and *New York Times*, was brought in to run the news network.) Discovery+ is also part of the same family.

At the time of writing this book, the division had not formally taken place. Not even the final names of the new companies had been confirmed, so predicting outcomes is somewhat tricky. One thing is clear though. Both have to deliver digitally if their parent companies and the executives who run them are to have a chance of surviving the streaming wars.[28]

Playing catch-up

It was not just Warner Bros Discovery that was struggling to make an impact in the new world of streaming. Other traditional studios that had to build tech infrastructure include Lionsgate, with its Lionsgate+ service. It barely landed a blow.

Ultimately, pretty much every meaningful media company in the UK and the US now has a DTC streaming service. Peacock was launched by NBC in July 2020. It boasts popular comedies like *Brooklyn Nine-Nine*, reality series like *Below Deck Down Under* and mega-hit movies like the Harry Potter collection in its catalogue. There are even 'live' channels, similar to the offerings from Samsung TV+ and Prime Video Channels but based on NBC content.

While it doesn't get the audience of other streamers, the bosses at NBC clearly realized, rightly, that they were fighting a losing battle in hoping younger viewers would keep paying for the traditional packages that contained their work and so launched Peacock. (The name is inspired by the bird that serves as the NBC logo.)

Fox, owned by the Murdoch family, has opted for a somewhat different approach. It has stuck with Tubi, a FAST streaming service as described above. It did this largely to maintain the importance of its various channels and its value to cable companies. However, things are likely to change as in February 2025 CEO Lachlan Murdoch told investors that it planned to a launch a service 'holistic of all of our content of sports and news' by the end of the year.

Underrated BBC iPlayer

The BBC does not have the Hollywood glamour that Netflix or Apple TV+ have, yet it is one of the pioneers of streaming. The iPlayer was rolled out in December 2007, the same year Netflix premiered its streaming offering, and has become more and more integral to Britain's national broadcaster, which has become increasingly focused on putting original work on there. For instance, when England international footballer Lucy Bronze chose to go public about her diagnosis of autism and ADHD in March 2025, she did so via a six-minute clip that was not aired on linear television but published on the iPlayer.[29]

Overall, the iPlayer is a very high-quality product, providing access to both linear TV and a whole range of content in its extensive catalogue. Streaming services from rivals like ITV and Channel 4 have long trailed it in terms of how well they actually work. (Using ITVX was, for years, a notoriously painful experience.)

The iPlayer also has the advantage, an unfair one according to many of its rivals, of its funding model, which is centred on the licence fee. Costing £174.50 per year in 2025, the licence fee has to be paid by anyone in the UK who wants to watch live TV, BBC or otherwise. It means the iPlayer is pretty much the only streamer to be both ad-free (save for skippable promos of other BBC content) and subscription-free.

Death of the cinema?

Did the increase in the number of streaming services and the habit changes this prompted cause the demise of the cinema? The kings and queens of Hollywood and their courtiers manning those picket lines during the 2023 strikes certainly think so. Common sense dictates that it must have had some effect, and there are plenty of figures to back up the argument too.

The UK Cinema Association, which represents around 90 per cent of UK cinemas, says that in the 21st century there have generally been 150 million or more cinema admissions per year. It registered 173.5 million visits in 2009. That number stayed fairly constant for a decade, with attendances at 176.1 million in 2019. It's not worth taking 2020 and 2021 into consideration (because of Covid-19 lockdowns) but in 2022, the number of admissions was 117.3 million.[30] That is a very sharp fall from just three years before. It recovered a bit in 2023 to reach 123.6 million, but that is still a significant way off the 2019 number.[31]

A similar pattern can be observed across the Atlantic too. In the US and Canada there were 1.225 billion cinema tickets sold in 1995 and 1.224 billion sold in 2019. In 2023, that number was 829 *million*.[32]

In the US, frequent cinema attenders – those who go on multiple occasions per month – have got back into their routine. Media research group Kagan found that frequent movie attendance was back up to 22 per cent of internet adults in 2022, a complete recovery to pre-pandemic levels. What had taken the hit was the infrequent attendees. The non-dedicated cinema goers got out of the habit, dragging down numbers overall. www.spglobal.com/market-intelligence/en/news-insights/research/frequent-us-movie-goers-are-back-infrequent-attendees-not-so-much

Perfect storm for cinemas

We will never quite know exactly what was behind the decline in cinema attendances, but streaming and the pandemic were undoubtedly a big part of it. Streaming was starting to gain traction and then the pandemic prompted a permanent change in viewing habits. Covid-19 forced the closure of cinemas during lockdown (some of which never reopened). During the time cinemas were closed, major movies were released via streaming platforms and everybody got used to the idea that they didn't need to leave their home in order to see the latest Hollywood hits.

As is so often the case, the most likely answer to why cinema attendance has fallen is a combination of factors – streaming, the pandemic and others. These include the growing financial strain people found themselves in, meaning they had less disposable income for treats like an increasingly expensive trip to the movies. If you were already cutting into your tightly stretched monthly budget for a streaming service, why would you spend even more to go to watch the same or similar work at a cinema?

Some releases did defy the trend. For example, the 'Barbenheimer' phenomenon, when *Barbie* and *Oppenheimer* were both released at the same time in the summer of 2023, driving huge numbers to the cinema. Blockbusters like *Avengers: Endgame* similarly drew big crowds when it arrived in 2019. These are a few standout moments, though, not regular occurrences.

As for the new entrants to the streaming space, by buying rights and commissioning their own movies, Big Tech stopped the traditional studios from being the only players in Hollywood. They were highly motivated to have people watch the movies they had bought through platforms and hardware they own. Sometimes this meant they could rescue work they had invested heavily in. After bombing at the cinema, Amazon almost immediately put 2024 Christmas blockbuster *Red One*, starring Dwayne 'The Rock' Johnson and Harry Evans, on to Prime Video.[33]

Previously, we used to have to wait months, or even years, for a movie to make it from theatres to the TV.

During the pandemic, Tom Hanks got into a bit of trouble for sticking up for the cinema. In an interview with *The Guardian*, he commented that it was 'an absolute heartbreak' that his movie *Greyhound* could not be released on the big screen because of Covid restrictions. 'I don't mean to make angry my Apple overlords, but there is a difference in picture and sound quality that goes along with [switching from the cinema to TV],' the Hollywood A-Lister said.[34] Shortly after, he had to walk his comment backs, presumably after receiving a stern telling off from his aforementioned overlords in Cupertino.[35]

Despite all this, the streamers are not willing, or indeed able, to totally ignore theatrical releases. A notable number of movies from tech companies are spending at least some time in cinemas. This is partly because awards tend to have certain theatrical release requirements for a movie to be considered.

There also remains an element of prestige in having a film out in the cinema that tech firms, and the directors and actors they employ, seem to like. Tom Hanks's comments made that abundantly clear. Top executives at tech firms have also shown a predilection for attending glitzy Hollywood events, incentivizing them to play at least a part in the traditional system.

Seeing films on the silver screen can absolutely be a great experience, with everyone laughing or crying together at the same time. Events like 'Barbenheimer' and the Avengers release were as much about being part of a cultural moment as they were about seeing the actual film.

Social media was filled with gangs of pink-clad women going to see *Barbie* together. However, the direction of travel is clear, and most of the time people are happy to watch what they want, when they want, where they want.

The increased quality in TVs and home sound systems means that, for many, the drop-off in viewing quality when streaming at home does not affect their enjoyment. Indeed, it can actually be a far more pleasant experience than watching a movie surrounded by random people chatting and coughing while eating overpriced snacks.

And who is it that makes and sells the TVs and other accessories with which you can build a good setup? It is, of course, the likes of Samsung with their big TVs and soundbars and fridges, Amazon with the Fire Stick and Apple with its TV box.

Netflix may have been the frontrunner, but now there are so many options providing a lot of reasons to stay on the sofa. These companies created more content, more services and more hardware than we had ever previously had access to. They ultimately gave us all more reasons not to head to the local multiplex. Instead, we gave the streaming services our money, and cinemas, blown out by a host of new entrants to the entertainment industry, undoubtedly became a victim of the streaming wars.

03
Streaming sports
A good deal for fans or a sign of things to come?

As camera swoops in on the teams' logos, the voice-over announces: 'Exclusively live from the City Ground, Nottingham Forest vs Liverpool'. The music rises to a crescendo – 'Here we go, here we go, this is it!' It is Sunday 16 August 1992, and football had just changed forever. The first games in the Premier League, the newly branded top-tier of English football, had kicked off the day before. However, this clash – 'Clough against Souness in the battle of the giants' – was the first of them to be shown on live television. The first Super Sunday.[1]

Sky promised fans 'a whole new ball game' when the Premier League launched. They certainly delivered, both for the good and to the detriment of the sport. In the decades after the dramatic start that Sunday afternoon, the Premier League became a crucial media product around the world, with money pouring into the game. People bought subscriptions to cable and satellite services as they became the only way to consistently watch Premier League football and other top-tier live sport.

While providing access to epic original series matters to broadcasters of all kinds, live sports rights have come to be among the most valuable assets they can own. The nature of these rights packages means they only periodically come to market and there is often a highly competitive bidding war when this happens. Knowing they are in demand, the various leagues are only too happy to keep raising their prices. If you want to show them, you need to pay up.

Sky Sports and TNT Sports acquired the Premier League rights for £6.7 billion in a deal that starts with the 2025/26 season for a four-year period. For the first time, the deal meant every game would be shown in the UK, outside of those kicking off at 3 pm on a Saturday afternoon. Black-out regulations mean those games are not allowed to be broadcast in the UK. The deal allowed Sky to show a minimum of 215 matches live, including all 10 of the games played on the season's final day. Sky and TNT were not alone in feeling the need to splash out for such an agreement. According to S&P Global, sports rights were estimated to cost $14.64 billion in 2015. By 2027, this number is expected to hit $35 billion.[2]

The growing cost of rights is not the only change in the sports media business. Having once been the buccaneering changemakers, Sky and others found themselves challenged by the streamers. The disruptors were at risk of being the disrupted. TV companies held on to sports rights in a desperate to bid to fight off the newcomers.

Why sport matters to streamers

Over time, the streaming services realized, as their cable predecessors had years before, that live sport is a key differentiator and a driver of ad revenue, customer sign-ups and retention. Slowly but surely, the newcomers started to enter the fray, fundamentally changing the game once again.

Advertising

The world of sports broadcasting looks very different to how it did back in 1992 when the Premier League launched, with many more players involved both on the pitch and in the media. This growth has had a profound impact. 'The more platforms that are being introduced and picking up rights… is adding to the fragmentation of the market,' explains Danni Moore, Senior Analyst at Ampere Analysis. These new outlets are keen to get involved because sport is one of the only things guaranteed to get lots of people watching all at the same time. The famous 'Sergio Agueeroooo' moment, Spurs' last-gasp win in the 2019 Champions League semi-final and Rory McIlroy finally completing golf's career grand slam would just not be the same watching on catch-up. As the 2022 marketing campaign from Sky Sports framed things: 'It's only live once.'

One senior industry insider put it to me: 'Sport is one of, if not the, last pieces of appointment viewing so it often guarantees people will tune in and watch.' Daniel

Harraghy, Research Manager at Ampere Analysis, endorses this view, explaining: 'Sport is one of the biggest, if not the only, appointment to view type of content now, where you're going to get sometimes millions of eyeballs in one place at a given time.'

Consequently, 'it's really valuable to advertisers,' says Harraghy. Cable TV has understood this value for a long time. That original Super Sunday introduction included branding from both Ford and Fosters. The money began to move as streamers get more of this content and work to build out the advertising side of their businesses.

'A lot of advertisers are opting to go to streaming,' Michael Frank, Principal Analyst – Sports Media analysis firm Omdia, explains. 'I think a lot of the big marketers know where this is going, and I think a lot of their advertising budgets are being redirected to streaming.'

Ultimately, 'if you want to have a well-rounded plan to get your product out to the masses, I think you have no choice but to move a large chunk of your budget over to a streaming service, where the eyeballs are,' he says. 'It's the age-old advertising model, and it's just part of the realm of media and watching the sport,' Frank notes. 'The sports teams and the streamers are happy about getting an influx of revenue from major blue chip sponsors.'

Customers might not like advertising in other types of content they stream, but the fact that this is an 'age-old' model in sports means fans are much more accepting. There are natural stoppages in action that provide space for advertising without the main event being disrupted. This is particularly true in US sports such as American football or baseball. Football has its half-time breaks too. 'I think it's just a very natural fit for those ad tiers,' says Harraghy. Ultimately, sports fans are just used to seeing ads and advertisers are used to being part of those broadcasts.

Sign-ups

Acquiring sports rights is not just about ad revenue though. It helps drive new signs-ups too. Sport has a dedicated and hugely loyal fanbase, so whoever picks up those rights knows that a whole chunk of people are going to have to use their service on a regular basis. If you are the place where fans can access their favourite team and league, they have little choice but to go there.

In a comment posted in June 2024, S&P analysts noted the increasing price of sports rights and how important acquiring them was for streamers. In one example, the authors explained how 'from its inception, Paramount+ has had sports as an integral part of its content offering because it provides access to all of the sports content that is on the CBS Network.'[3] It also has a number of games from the men's Champions League – Europe's premier club football competition – exclusively on the platform. Peacock has also made showing the Premier League – the top tier of English football – a central part of its offering. In America, if you want to watch those games then you have to subscribe to those services.

It is similar in Australia: if you want to watch the weekly Premier League action Down Under, there was no choice but to subscribe to Optus Sport, which originally acquired the rights to the games in 2016. In 2021, it extended the deal for a further six years, paying $58 million per season to be able to show all 380 games in the league on its streaming service. It beat off streamers Kayo, Stan and Prime Video to win the auction.[4] Stan did get the rights to show the Champions League Down Under, and eventually picked up the Premier League rights too as Optus Sport was shut down in 2025.[5]

As well as getting the original sign-up, sport also means customers have to keep their subscriptions for longer. You cannot binge watch a 10-month-long football season like you can a 10-episode season of a TV show. The streamers also hope that the sport brings you in but then you find other things in their catalogue that you want to watch and so stick around. The dedicated sport streamers also tend to have a package of rights, so you stay all year long. Stan also shows rugby and tennis too, for example.

Peacock is a prime example of how sport can drive sign-ups. As discussed in the previous chapter, it has a collection of hugely popular shows within its catalogue, but it also became the new home of Premier League football in the US. Almost every match is shown there every week of the season. Things didn't stop with that. Parent company NBC grabbed the rights to NFL matches, college football and some basketball and put some of those on the platform too. This included making Peacock the only place to watch the Miami Dolphins take on the Kansas City Chiefs in their January 2024 Wildcard playoff match. The game drew in 2.8 million sign-ups over the course of the NFL's Wildcard Weekend, with an average of 23 million people watching the clash. The Peacock broadcast took up 30 per cent of internet traffic at the time.[6]

NBC also had the rights to the Paris 2024 Summer Olympics and they went all-in with their streaming service. Not only did Peacock show every event live, it launched Gold Zone. A play on the NFL institution Red Zone which broadcasts every key moment and touchdown, Gold Zone showed each Gold Medal being won. Red Zone host Scott Hanson was even drafted in to lead the coverage on its Olympic sibling.

It was all a huge investment from NBC but it ended up paying off. Users streamed 23.5 billion minutes of action from Paris 2024 on Peacock, by far the largest American viewership from any Games.[7] Nielsen reported that Peacock scored 2.1 per cent of all views in August 2024, a huge win in a country that is usually apathetic towards the Olympics, at best.[8]

Analysts S&P acknowledged that Peacock it is 'far smaller' than lots of the rival services it is competing against, but 'has employed sports as a key part of its growth strategy'.[9]

Mixed it up

The reality is that, just like cable TV, sports streamers use a mix of both subscriptions and advertising in order to maximize revenue. There are very few services that show sport for free and almost all show advertising, with the BBC and its iPlayer among the small group of exceptions. As noted earlier, Peacock had ads on the platform, right from the start. The combination of ads and popular sport helped Comcast/NBC Universal make income from streaming 17 per cent of its total ad revenue in 2023. That is a crucial shift for mid- and long-term sustainability and there can be little doubt that the ability for brands to advertise against premium sports content on the platform helped achieve it.

One thing to note across all these services is that even those who pay for ad-free tiers of streamers are served adverts when live sport is on; i.e. even if you're paying for a top-tier Netflix subscription, you will still see adverts during the NFL games it shows. There were ads in the Premier League matches on Amazon Prime Video long before they were rolled out as part of a standard subscription too.

It is worth reiterating that while the introduction of general ad tiers has provoked some backlash from viewers, having them in sport has proved to be almost entirely uncontroversial. Viewers are so used to seeing advertising during breaks in play that it barely raises an eyebrow. It makes sport even more valuable to streamers, as it is a great way to grow their ad inventory.

A difficult game to play

So the incentives are very much there for the streaming companies to get involved in sport. However, as Netflix demonstrated, moving into this area, and live events more broadly, is not entirely straightforward. The costs of the rights are high. Production costs are steep. The pressure to deliver is huge. Broadcasters absolutely cannot get it wrong when it comes to live sport as fans will make their displeasure very clear indeed.

Amazon found this out when it first showed Premier League football in the UK in December 2019. Its move led to a record number of sign-ups for Prime Video.[10] However, it also led to a number of, almost widespread, complaints. Fans griped about buffering issues, the quality of the stream and delays.[11] I remember hiding my phone away during games because it was possible that by looking at X (formerly Twitter) or getting notifications from a sports scores app I would find out about a goal or similarly important event in the match before I had actually seen it on the stream. Far from ideal. As Adam Dalrymple, a media analyst formerly at Enders Analysis, put it: 'Sports is incredible because it's a genre of content that has incredible fandom, but the flip side of that is, if you get the broadcast wrong, you will turn all of those fans against your brand.'

He explained that 'either streamers entering sports, or broadcasters shifting to streaming, are on their own learning journey on this point, but it's about whether they can make the investments upfront to ensure that there isn't a sort of catastrophic cutout during a major game'.

Things definitely improved as time went on. The varying reliability of household internet connections means a provider can never guarantee that everyone will see the same goal or terrible refereeing decision at exactly the same moment, but Prime Video regularly showed Premier League and Champions League football, not to mention live Thursday night NFL matches in the US, with relatively little complaint. It's as if we've all got used to the new streaming reality.

Netflix starts throwing punches

Perhaps the pinnacle, or nadir, depending on your approach to sporting integrity, of this new sports streaming world was when Jake Paul took the ring with Mike Tyson live on Netflix in November 2024. The company claims that the fight was the most streamed sporting event in history with 108 million viewers tuning in, with 65 million concurrent streams around the world at one point.

Despite those numbers, the sight of a 58-year-old former world champion with criminal convictions taking on a YouTuber-turned-fighter was a fairly grim one for boxing purists. The bout with real sporting significance was the penultimate one on the card, as Katie Taylor beat Amanda Serrano for the women's undisputed super lightweight world title. Netflix said that fight, which it branded the co-main event, drew in a global audience of 74 million. This, according to the streamer, made it the most-watched professional women's sports event in US history,[12] demonstrating the boost an event can get when broadcast on a global platform. Netflix was the host once again when the bitter rivals exchanged blows for a third time on 11 July 2025 at Madison Square Garden in New York.

Despite the improvements from others, the tech issues that Netflix had encountered at other live events reared their head again, with thousands of viewers complaining about their stream buffering. Netflix can get the numbers, but until the technology is stable during its most popular events, it will never be a major sporting player. Indeed, things actually got to the point where people tuned into these major Netflix events simply to see if they are actually going to work.

In fairness, watching the Christmas Day 2024 NFL games was generally considered a far smoother experience. No doubt some people endured buffering issues, but there was no widespread outrage as there had been at other events. It even seemed to be fine when Beyoncé took to the stage of a Super Bowl-like halftime show. Netflix worked with the NFL on the product and under the terms of the deal with the NFL Network, the games, and Beyoncé's performance, would expire in the US three hours after they took place. This is because the network was going to re-air them. Such an arrangement may well be a sign of how things will develop in the future, with this with the live broadcast experience collaborating with streaming services as part of a global deal.[13]

In DAZN

Netflix might have dabbled in boxing, but the undisputed streaming home of that sport is DAZN. While the service has added a variety of offerings to its roster, it is best known for its involvement in fights sports, and boxing in particular. Founder Sir Len Blavatnik was even featured in *Sports Illustrated*'s Most Influential People in Boxing list in 2024. The magazine described DAZN as 'the go-to platform for fight fans'.[14]

Sir Len is a fascinating, controversial character. A Jewish émigré born in Ukraine in 1957, he returned to the

Soviet Union as it fell apart and got rich as the assets were privatized.

DAZN, sometimes referred to as the Netflix of sport, was founded in 2015. By January 2025, Sir Len had poured $6.7 billion into the streamer as it continually failed to make a profit. The company generated revenue of $2.9 billion in 2024, but lost $1.4 billion overall that year. In a January 2024 interview with the *Financial Times*, CEO Shay Segev said that his company was profitable in most of the top 10 countries in which it was available. 'We see a clear path to create massive shareholder value. We're going to be the global home for sport. If you are a sports fan anywhere in the world, you are going to know DAZN,' he insisted.[15]

Unlike Apple, and on some occasions Netflix, DAZN picks up different rights in different countries, instead of insisting on showing everything globally. For instance, in 2024 it acquired the rights to be the exclusive Champions League broadcaster in New Zealand for the following three seasons.[16] It had already purchased the same rights in Canada back in 2018, an arrangement it extended for three years in 2024.[17]

While this makes using DAZN quite a disjointed experience, one dependent on which country you are in, it does help the firm enter markets more easily, picking up rights as and when it can. It also means that it can show sports where the interest might be limited to a certain country, such as the purchase of the Aussie Football League (Aussie rules football) and National Rugby League (NRL) (rugby league) rights in Australia in May 2025.[18] These two competitions are wildly popular Down Under but make a limited impact in most other countries.

Perhaps most eye catching was that DAZN became the home of the Saudi Pro League just as iconic figures such as Cristiano Ronaldo arrived, bringing with him attention and the need for a broadcast partner.

The service has been a big supporter of women's football too, showing Champions League and domestic fixtures from places like Spain and Italy. It made many of these broadcasts available free for anyone who created an account.

DAZN also serves as a platform in a similar manner to Prime Video. In the UK, using it is the only way to buy the NFL Gamepass. There is a similar product available for the fourth tier of English football, the National League, as well. In addition to asking for a monthly subscription, some of the highest-profile boxing bouts require an extra payment, no different to Pay Per View on a cable network.

It hasn't all been plain sailing though. As well as the struggles to turn a profit, DAZN was engaged in an ugly dispute with France's top football league, Ligue 1. In 2025, a deal between the broadcaster and the league that was due to last until 2029 fell apart.

There were stipulations within the contract that either the league, DAZN or fellow broadcaster beIN Sports could back out after the second season. Given DAZN was scheduled to show at matches a week, this was a crushing blow for a league that had already cycled through four major broadcast partners in five years. According to information published by Enders Analysis, there were allegations that Ligue 1's parent company LFP was not living up to its contractual marketing obligation and, more seriously, was not doing enough to stop its feed being pirated. Things descended into litigation and mediation. Ligue 1 decided to stop working with third-party broadcasters and instead create its own DTC offerings, something all other top European leagues have done to some degree, with the notable exception of England's Premier League.[19,20]

The scuffle with Ligue 1 aside, work from Ampere Analysis showed that DAZN was set to be the top investor in sports rights. Having splashed out $1 billion to show the FIFA Club World Cup, it accounted for one-third of streaming expenditure that year.[21] Its development from specialist broadcast to one with a range of assets and payment models makes DAZN a fascinating participant in the streaming wars.

A fair deal for fans?

As the amount of sport available via streamers increases, customers will be able to move away from their traditional satellite and cable packages and get the content solely via streaming services – so-called cord cutting. 'I think the pay TV market is losing,' said Omdia's Michael Frank. 'We know that subscribers are going away from pay TV… And as more and more sports rights go to streaming, you're going to see people and subscribe rates go lower and lower.' Bad news for linear broadcasters. Great for streamers.

While sports leagues and streaming services may have benefited from the move away from being purely on linear networks, the question remains as to whether it is actually good for fans. Yes, they may no longer have to pay for an expensive cable bundle, but the combined cost of the various other services soon adds up.

There can be no doubt that you need more services than ever before to follow popular sports. This is an increasingly regular complaint towards the streaming industry as a whole, but it is a particularly pernicious issue for sports fans. It is quite a task for them to keep up with where their team is being shown and when – broadcasters of all kinds generally get to set the time slots matches take place in and pick which teams play in them.

In the UK, in order to watch every televised football match featuring an English league side during the 2024/25 season, a supporter needed to have Sky Sports (or Now), TNT (or Discovery+) and Prime Video, which showed one Champions League match each Tuesday and had two blocks of Premier League matches in December. While Amazon decided to give up on the Premier League after that season, it retained its Champions League rights, alongside TNT, which also has FA Cup rights, but has a deal with the BBC to keep some fixtures free-to-air.

It get seven more expensive if you want the action from other major football competitions from around the world. Ampere Analysis found that the cost of getting access to 15 top competitions, including the English Premier League, Football League and FA Cup, Italy's Serie A, Spain's La Liga, Germany's Bundesliga, the Champions League and America's MLS, went up from £89.23 per month to £140.21 between the 2019/20 and 2024/25 seasons. That's a hike of 57 per cent to access all eight of the services required to tune in.

It's worth noting though that streaming can actually help limit price rises. If you are happy to access matches via the Now and Discovery+ streaming services instead of having the full fat versions of Sky and TNT, for instance, that price rise was only 5 per cent over the same period.[22]

Things are even more complicated in the US. Six TV and cable networks and a multitude of streamers provide access to the NFL, the country's most popular sport. It is a similarly convoluted situation for other popular sports like basketball and baseball too. American sports tend to have their own dedicated streaming services, such as MLB TV for baseball, as well as the television broadcasts. What is available is often dictated by someone's location within the States. For instance, a New York Yankees fan in New York itself needs to access local TV to watch the games. They will not be available via MLB TV because of geo-restrictions. This all means lots of access for fans, but also lots of expense and time spent checking where they can actually watch things.

As with elsewhere in streaming, not having a monopoly should be a good thing for consumers. It should mean service providers need to consistently raise standards and offer an acceptable price in order to compete against one another. However, this doesn't quite work in sports broadcasting because the same matches are not shown in multiple places in one country. It's not like deciding whether to get a cappuccino at one cafe or the other! If you want to watch the game legally, you have to hand over your cash to the one place where you can do so. Consequently, most of the benefits of the rising competition between broadcasters, including the introduction of streamers, have gone to the leagues. They can play the broadcasters against each other, driving up the value of the rights, the cost of which is inevitably passed back to the fans.

Growing sports

Another reason the broadcasters still hold on to plenty of power is the fact that their money fuels the growth of the sports. This means that they can dictate the schedule of matches and other important factors knowing the leagues generally need to play ball. The intervention of major broadcasters can be blessing and a curse for growing sports, such as women's football in the UK.

For the 2024/25 season, a deal was struck so that the Women's Super League, the top tier of women's domestic football in England, would be shown across Sky Sports, the BBC and YouTube in its home country. This meant all the games were available to watch, a significant number for free. It was a great way to develop a competition that had got increasingly popular in the wake of England's Lionesses winning Euro 2021 (actually played in 2022) on home soil.

The shift to YouTube was also a major technical improvement on the almost unusable FA Player, the free streaming service that had previously hosted non-televised WSL matches. As well as undoubtedly improving the fan experience, moving to the already popular and established YouTube also helped drive viewership up.[23] While organizations like the FA might wish to build their own platforms and control the presentation of a product, sometimes it is better to hand things over to those who really know what they are doing and already have an audience.

While a YouTube tie-up was positive for women's football in the England, the scheduling chosen by the broadcasters often led to fixture clashes with the men's team from the same club, presenting fans with a dilemma over what to watch and attend. Commenting ahead of one match at which his side's match coincided with one from their male counterparts, ex-Tottenham Hotspur Women's manager Robert Vilahamn acknowledged the different factors at play. 'I want the Tottenham fans to support the women and men's team and be able to go watch both games,' he said. 'I know that we work really hard to do that, but then you have the TV companies that also need to make sure they can dictate it, because they provide a lot of money into the business as well, so you need to respect that.'

As well as YouTube, Netflix has decided to make a big intervention in women's football. It secured the exclusive US rights to the 2027 and 2031 FIFI Women's World Cup, marking the first time the tournament went to a streamer. In announcing this deal, Netflix Chief Content Officer Bela Bajaria referred to her company's 'record-breaking success with Amanda Serrano vs Katie Taylor', the fight that was the co-headline bout with Jack Paul and Mike Tyson. She said the viewership of that fight 'demonstrated the massive appetite for women's sports'. FIFA boss

Gianni Infantino declared the deal 'a landmark moment for sports media rights' and insisted 'Netflix has shown a very strong level of commitment to growing women's football'.[24]

Being on something as widely used as YouTube clearly helped the WSL become more accessible, so the confidence around the introduction of Netflix is understandable. In reality though, a sport such as women's football needs to appear regularly on a popular, widely accessible linear channel in order to really grow. It is notable that when a new entity took over the top two tiers of English women's football in 2024, it struck a new media deal with Sky and the BBC, as well as keeping YouTube. This landmark deal allowed for all 132 WSL matches to be shown from the 2025/26 season, proving that the leagues still value the heft the old-fashioned broadcasters have.[25]

Niche sports

One huge benefit of streaming to sports fans is that it allows followers of more niche sports to be able to watch them live far more easily. A good example of this is handball. Hugely popular in many European countries, it is, nonetheless, not at the level of some other sports such as football. However, both the most popular domestic league, Germany's Bundesliga, and the continent-wide Champions League are available through their own dedicated streaming services. This means that instead of negotiating with TV companies, who may be less enthused by a sport than the fans undoubtedly are, the competitions can bring the action straight to those who want to watch it.

The service streaming the Champions League, EHFTV, provides a good viewer experience and means fans almost anywhere in the world can tune in. A dedicated supporter of the German team Magdeburg is probably very happy to be able to watch them live in both domestic and European competition, wherever they are.

As ever though, there are pros and cons to this approach. As the FA learned with the WSL, a standalone service might theoretically make a sports league more accessible as there is only one place to go, but that does not necessarily aid discoverability. You're unlikely to stumble upon Champions League Handball if you live in the UK and it is unlikely you would have found a WSL game on the FA Player without knowing where to look. On YouTube or mainstream TV, discoverability shoots up.

In reality, smaller or up-and-coming sports and broadcasters need one another. 'Streamers and legacy channels and broadcast channels and networks need content,' explains Omdia's Michael Frank. He notes the battle for the rights to TGL golf set up by Rory McIlroy and Tiger Woods, as well as women's volleyball, 'pickleball leagues and major league table tennis and Professional Bull Riding League'.

It is a 'symbiotic relationship between media rights and sports leagues', according to Frank. Not only do the new leagues provide programming, they 'need a partner, like a major television network, or… a streamer, to get their content to the masses'. Channels need to fill their schedules, sports need exposure. And, really, which outlet *wouldn't* want to be known as the home of professional bull riding?

Whatever the sport, showing it has always been crucial to cable and satellite companies. Now streamers are seeing it as a vital part of their offering too, adding to the amount of competition there is when the broadcasts rights become available.

Bending the rules

So far this chapter has looked at purely legal sports streaming – services buying rights and viewers in the appropriate country paying to watch the broadcast of them. As I write, another screen in front of me is streaming an England vs Australia cricket match via Sky's Now service. Sky has paid for the rights in the UK, I've paid for Now in the UK. Simple. However, there is another side to all this that veers between the grey market and pure illegality.

Thanks to Virtual Private Networks (VPNs), committed fans are now able to find coverage of almost any sport they want online, from anywhere in the world, sometimes even for free. VPNs are perfectly legal software that, at the most basic level, let you change the location of your computer. They add an extra layer of security for the user particularly if they are out and about using an unsecured network, hence why they are absolutely legitimate.

A VPN also means that Brits can set their computer's IP address, a device's digital location, to the USA and find a way to sign up to Peacock or another American service discussed above. (Doing so usually requires an American bank card and address, but these are not insurmountable barriers.)

Equally, VPNs allow Americans to put their computer in the UK, at least virtually, and watch the BBC for free. In fact, with a quick click of a button you can view content from anywhere in the world. This, as I say, is something of a grey area. Both VPNs and the service being accessed are completely legal, but the person viewing is not actually meant to be accessing them from the country they are physically in.

Then there is the pure black market of sports streaming. Users can scramble around the web to find a less than legitimate site providing a feed of whatever game they want to watch. These sites are usually peppered with annoying ads and potentially nasty computer viruses, but they can also give the viewer access to a game they want to take in at no cost. On top of this, there are various hardware boxes and software services that provide access to TV channels from around the globe, no VPN required, for relatively little money. Law enforcement is generally more interested in wiping out the existence of the service itself than trying to hunt done the various people using it. It's a bit like a game of whack-a-mole. If the authorities take down one site or service provider, another is likely to spring up in fairly short order. Fans want to watch and there will always be those prepared to run the legal risk and facilitate this and earn some money from the accompanying adverts.

The cost of cable channels and TV services encourages a search for alternatives. VPNs don't cost very much, many illegal streams are free and other options are certainly cheaper than a Sky or a cable subscription. The number of subscriptions required has a similar effect. It is tempting to just pay for illegal service that makes everything available in one place and not a handful of the legitimate ones, especially if doing so actually gets you access to more matches. Both broadcasters and the leagues need to be careful that they do not kill the golden goose by asking for more and more, and pushing fans to look elsewhere.

Bin the black-out

As well as the cost of watching sport, the way some of them operate makes such behaviour rather tempting. The clearest example of this is in the UK, where no broadcaster of any kind is allowed to show football played between 2.45 pm and 5.15 pm on a Saturday afternoon.

These antiquated black-out rules were designed in a pre-streaming era and are a particular bugbear of mine; one, I should add, that I share with Prince William. In an interview, he told *The Sun* that the rule is 'irritating' as it limits the number of games he can watch his beloved Aston Villa play. 'It's really annoying that we still can't watch our own team's Premier League match in this country but you can go abroad and watch any game any time,' the heir to the throne said.[26]

The origins of the regulations that frustrate me, the future King and a host of other British football fans are understandable. The black-out was implemented to encourage people to go to matches instead of staying at home and watching them on TV. There were particular concerns that local lower league teams, a crucial part of the British football pyramid and their communities, would lose support and vital revenue if people could just watch the big boys from the comfort of their sofa.

However, in the streaming era the black-out has become nonsensical. Determined fans are going to find a way to watch a game whether or not it is being broadcast on television in their country or elsewhere. (And they are all on somewhere.)

It could be argued the rules are limiting the growth of the women's game too. The WSL was not allowed to ignore the black-out when it negotiated its broadcast deal in 2025.[27] Fixture clashes with their brother teams might be more easily avoided if one of the sides could be televised on a Saturday afternoon and more potential fans would get to watch women's football.

All in all, it would be far better to allow broadcasters to show the game in the traditional 3 pm Saturday afternoon slot adored by fans. Given that all the Premier League football rights holders have streaming platforms, there is no technical impediment to showing multiple games at the same time. Broadcasters are no longer limited by the number of linear channels they have.

Another alternative to the black-out could be the game-pass model, as used in the US. For one payment fans can watch the NFL games not being broadcast on TV. Basketball and baseball operate similar models. Whatever the solution, it is time for the black-out to be binned.

Wrong Venu, wrong time

As streaming became increasingly prominent, so did conversations about 'rebundling' – a group of services coming together to offer a package that includes them all in one monthly price. Given the market fragmentation and fan frustration, not to mention the ever-increasing subscription costs, this conversation was particularly relevant in sports.

In February 2024, ESPN, Fox and what was then still Warner Bros Discovery – big broadcasting beasts who between them held many of the major sports rights in the US – announced they were joining together and launching a streaming product called Venu. As part of this so-called 'sports skinny bundle', viewers would get access to a range of the broadcasters' TV and streaming products such as ESPN, ESPN+, Fox Sports 1 and 2 and TNT in a single service. The intention was to make things simpler and more cost-effective to users, while allowing the companies to move further into the DTC market. (Fox had been particularly reticent about doing so.[28])

The project had its own CEO, Pete Distard, and was set to be priced at $42.99.[29] Distard declared that his team was 'singularly focused on delivering a best-in-class product for our target audience'.

While the three partners did not have the rights to every piece of sporting action, there was a lot there that would appeal to fans, such as NFL, basketball, Major League Baseball, ice hockey, tennis and football. All in all, it was a tempting offer. There was some cynicism though. As the sports media columnist Andrew Marchand pointedly put things for *The Athletic*: 'One day, the brilliant TV executives are all going to unite and put their programming under one roof. It will solve all your sports viewing problems. They will call it cable.'[30]

As you might imagine, creating and launching a product such as Venu was not straightforward. Some of the channels set to be available through it could be accessed via other streaming services, meaning it might not work out to be all that enticing to viewers who wanted to watch a range of sports. As Marchand noted:

> There probably are some sports fans who would like to save a little money with this arrangement, but it is hard to believe there are a lot. You already can watch nearly everything that this trio offers through places like YouTube TV for around $70 and change per month. If you want this option, it is already available, with even more channels to boot.

All the while, ESPN continued to develop its own standalone streaming product codenamed 'Flagship'. It was intended to be more expensive and give viewers the full ESPN experience.

Most significant of all, rivals were (unsurprisingly) furious about Venu. Fubo, a company that offers a streaming service as an alternative to cable, took legal action against the new project. In August 2024, a US District Court of the Southern District of New York ruled in favour of

Fubo, agreeing that the arrival of Venu would 'at least tend to lessen competition in the Live Pay TV Market'.[31]

This eventually saw the entire plan for Venu scrapped in January 2025, without it ever going live. In a statement announcing the demise of the project, the three companies said that they had 'determined that it was best to meet the evolving demands of sports fans by focusing on existing products and distribution channels'.

ESPN's own plan for a DTC digital offering did come to fruition. In May 2025, the company announced that the much-anticipated Flagship was actually going to be called... ESPN. This, Chairman James Pitaro said in a press release, was because 'as we thought about the name, we kept returning to the fact that, across every generation, ESPN is the most trusted, loved and recognized name in sports, and that we should keep it simple and double down on the power of ESPN'.

This new app would bring subscribers to all of the company's linear properties, and there was also an offer allowing users to bundle the new app along with Disney+ and Hulu. Pitaro promised that the new service would 'be the ultimate sports destination for personalized experiences and features'.[32]

Sport and linear TV have a very co-dependent relationship. Cable is still a crucial part of most sports lovers' media diet and it is sport that largely brings meaningful audiences to cable. Streamers know this and want to get involved, increasing the amount of fragmentation and cost fans have to contend with. As Venu showed, solutions to this are not simple to find.

Without question, as time goes on, streaming services will pay to show more and more sport. Like it or not, fans are going have pay more to keep up with them across an expanding number of services.

04
Livestreaming
YouTube, Twitch and a darker side to streaming

When most people think of streaming, they think of the companies we've discussed so far in this book, the likes of Netflix and Disney. They think about services that allow them to watch on-demand content through an app via the internet. However, for many others, that is not at all what streaming is about. For them, it is a live conversation, an interactive broadcast. It is playing video games while dozens, hundreds or even thousands watch along while you do so. For a select few, it is a source of serious income.

Livestreaming is a hugely important part of the modern media environment, particularly for younger people. There are others, but the key platforms are Twitch and YouTube. They battle for supremacy in this field. Twitch has a real focus on video games and describes itself as 'the place where millions of people come together live every day to chat, interact and make their own entertainment together'.[1]

YouTube is likely more familiar to many readers. It introduced the livestreaming functionality in April 2011, although it had had some one-off live events before then.[2]

The top players

The biggest live stream creators, the individuals and groups broadcasting themselves, tend to operate across both of these platforms. As we will see, though, there have been attempts to tie down popular figures to one or the other with lucrative exclusivity deals.

Tyler Blevins, known to his fans as Ninja, is pretty much the biggest live stream creator of all. At the time of writing, he has over 19 million subscribers on Twitch and close to 24 million on YouTube. People spend hours watching him play *Fortnite* with fans posting in the chat room. There are plenty of others worth mentioning too, among them Kai Cenat, who has over 16 million followers on Twitch. He plays video games as well as pulling off stunts IRL (in real life).

To put some monetary value on all this, in 2023 Forbes estimated that Blevins was worth $10 million.[3] Perhaps we all should have spent more time playing video games instead of doing our homework after all…

Central to the success of these top livestreamers is their personality. They are engaging. Fans want to 'hang out' with them for hours at a time and support them financially. The most successful have built large communities, and businesses, around their presence online. As is the case across the creator economy, with the likes of

Instagram and TikTok, followers feel they really know their favourite livestreamers.

For many of us, it is hard to understand why anyone would want to spend large amounts of time watching someone else play video games, never mind giving them money in return for digital assets or just as a display of affection. The divide, it must be said, tends to be generational. Twitch parent-company Amazon says that 70 per cent of the service's users are aged between 18 and 34. It also claims that 65 per cent of users cannot be reached via other platforms,[4] explaining why the platform has a distinctive feel to it.

In 2023, Twitch averaged 105 million monthly visitors. That same year, 1.3 million minutes of content were watched on the platform.[5] By contrast, almost 100 *billion* hours were viewed on Netflix over the same 12-month period.[6] So livestreaming remains, in relative terms, small. That doesn't diminish its significance though. For those both broadcasting and watching, livestreaming is hugely important. Given this, it is no surprise that various social networks have also integrated live broadcasting capabilities into their apps. TikTok has put a particular emphasis on this, including adding monetization incentives for those who broadcast on the platform. Instagram has live stream functionality too.

As with the rest of the streaming, we're no longer talking about people just getting their content through computer monitors or their phones. Smart TVs mean that viewers are taking in these broadcasts in the same way they take in other forms of live TV.

That's not to dismiss the importance of smaller devices though. Research from Custom Market Insights (CMI) found that 'the increasing adoption of smartphones across the globe is driving the market'. This has been enhanced by the widespread roll-out of 5G internet, which means people can both stream live and watch live content on the go far more smoothly. CMI estimated that the entire market was worth $88.5 billion in 2023, rising to $104.9 billion the year after. By 2033, the analysts anticipate that the market will be worth a whopping $712.9 billion.[7]

In a similar vein, Enders Analysis research detailed that online video viewing was starting to move 'away from social media apps which pivoted to video like Instagram and Facebook, and towards video-first platforms like TikTok – which is also expanding its video lengths – and Twitch'.[8] The time, discipline and dedication required to be a successful livestreamer is significant, but it's now clear there is a significant audience and wealth to be found in return. Twitch undoubtedly has first mover advantage, but there are challengers, and not just YouTube.

Some other streaming and social media platforms, the world of livestreaming has its own unique culture and language. It is a world of 'subathons' – staying online for hours or days at a time to bring in new subscribers – and 'marathons' – staying online for a long time. Someone's entry into this universe may have started in their bedroom, but it can result in them being broadcast around the globe in real time, connecting them with all sorts of people along the way.

Twitch vs YouTube

I mentioned that the two key platforms when it comes to livestreaming are Twitch and YouTube. The former launched in 2011 as a place for gamers to show off their skills, and that is still very much at the core of what it is about.

In August 2014, Amazon bought Twitch for $970 million. Commenting at the time, Amazon founder and then CEO Jeff Bezos rightly said: 'Broadcasting and watching gameplay is a global phenomenon and Twitch has built a platform that brings together tens of millions of people who watch billions of games each month.'[9]

That phenomenon has only grown, as the 2023 stats revealed. While the service is undoubtedly popular, it doesn't actually generate a profit, a recurring theme across much of streaming and social media. A decade after that Amazon acquisition, *The Wall Street Journal* saw documents that revealed a slowdown in both growth and engagement. A few months before that, CEO Dan Clancy had said: 'I'll be blunt, we aren't profitable at this point.'[10]

There is an argument that Twitch is such a small part of the Amazon empire that the parent company is not all that worried about it. It is also true that the service is expensive to run owing to the computing power required to host livestreaming, making profits harder to come by even if the concept is popular.

It is now owned by one of the world's biggest companies, but Twitch was originally a spin-off of another product called Justin.tv. This no longer exists and had a much broader focus than video games. That original idea came from an attempt to make a 24/7 live documentary about one of the company's co-founders, Justin Kan.

Since being spun out of that original project, Twitch has built an integral role for itself within the gaming community. This led to others wanting a piece of the action.

YouTube goes into battle

After offering only small amounts of live content before, in April 2011 YouTube launched YouTube Live, declaring that 'the goal is to provide thousands of partners with the capacity to livestream from their own channels in the months ahead'.[11]

It is worth pausing here to consider once again the significance of YouTube more broadly in the context of streaming. It is essentially a social media platform that has been around for 20 years, yet it remains more relevant and popular, and just as groundbreaking, as when that first video from San Diego Zoo was posted there. In a space where popular services can seemingly evaporate overnight, this is a huge achievement.

YouTube's longevity is not just down to being part of Google. How the service rolled out livestreaming is another example of how it uses its generous revenue-sharing programme to attract creators.

YouTube clearly recognized it risked being left behind in the livestreaming space. And it had one huge weapon in its arsenal to allow it to make an impact – the Google advertising network, AdSense. Offering this functionality and an already popular platform was a compelling offer to creators thinking of getting into the livestreaming game.

With AdSense, livestreaming on YouTube can feature adverts, before and during the broadcast. Twitch also has advertising functionality too. Arguably, its Stream Display Ads (SDA) are more appropriate than YouTube's offering. As the stats above show, brands may also be able to reach viewers that they simply cannot get anywhere other than Twitch. All very appealing to marketers. However, Google's ad network is one of, if not the, most powerful in the world, making it extremely attempting for creators to join.

Golden handcuffs

As YouTube Live has developed, the battle between it and Twitch has become increasingly intense. One key element of this has been each company trying to ensure the best, most popular creators are tied down to their platform. This has involved hugely valuable contracts being offered in return for a creator operating in only one place.

In January 2020, YouTube signed three major players to exclusive deals: Lannan 'LazarBeam' Eacott, Elliott 'Muselk' Watkins and Rachell 'Valkyrae' Hofstetter. The latter had previously focused mostly on Twitch, where she had gained close to a million followers. (She had a similar amount on YouTube too.) Her move was hugely significant and negotiations were done by a major talent agency, UTA, which represents some of the biggest stars in the world.[12]

Burnout

The idea of sitting for hours playing video games, and being paid handsomely to do so, sounds like a dream come true to many people. And no doubt creators who partake in this kind of work have plenty of fun. Doing what they do is certainly more enjoyable than having a 'proper' job. However, that doesn't mean being a professional livestreamer is a totally carefree existence. They are expected to be online for hours at a time, only taking minimal breaks. Spend too long in the bathroom and you risk viewers getting bored and walking away. Don't broadcast consistently enough, and they will forget you exist.

People tend to start livestreaming at a relatively young age and the commitment required to really be successful can eventually lead to burnout. 'Around 200 viewers were when it started getting exhausting,' Stephen Flavall, who goes by the handle 'jorbs' on Twitch, told NPR in 2022. By that point he had grown to 2,000 concurrent viewers when he was broadcasting. He found that 'when that many people are asking you questions and telling you what to do, it becomes absolutely unmanageable'. He suffered from anxiety and symptoms akin to panic attacks.[13]

Thankfully, things improved and jorbs built a team to help him continue to grow his channel. Three years on from that interview he was still posting regularly and was up to 114,000 followers. What Flavall had to work through was not a new phenomenon though. Speaking in 2021, Imane Anys, known online as Pokimane, told *The Guardian*:

> Burnout is an incredibly real thing in gaming. A streamer sets their own work hours and it can be easy to fall into the trap of streaming 8 to 12 hours a day, seven days a week. It's frightening because people grind crazy long hours, and see results – hence why so many do it.

At the time, Pokimane was the most popular female streamer with 8.4 million subscribers.[14] By 2025, she was up to 9.4 million as she continued to put out a consistent stream of content.

Twitch is not all about gaming, and it is perhaps the more intimate styles of streams that can really take their toll. There is the Just Chatting category where people talk about their lives and engage in the chat room for hours at a time; sports where people watch games while broadcasting; and Pools, Hot Tubs, and Beaches, which can get a little bit more... streamy. (Those who want more of this type of content tend to head to another livestreamer – OnlyFans.) All in all, it is amazing how much of themselves people are willing to display to the world and for how long.

By May 2025, Emily (she doesn't publicly reveal her last name and appears as EmilyCC online) had been broadcasting live on Twitch continuously for over three years, having started testing the waters in 2016. It was a record, one that has generated a decent income even when the follower and viewer counts fell. The cameras stayed on while she slept and she carried a backpack with a camera attached when she went outside to walk her dog or do anything else. The only camera-free space in her life was her bathroom. The original motivation was to earn enough to buy a property but watching the channel you get the sense things had gone beyond that, that Emily was now undergoing an endurance test she wouldn't, or couldn't, give up.

Getting to the level of followers she had, over 325,000 as I write this in May 2025, had taken its toll. Emily suffers from migraines, depression and has basically lived in isolation since the cameras turned on. Even birthdays were spent on her own in her stream. A romantic life was impossible.[15] No wonder that some days she just hides under her bed covers away from the screens.

Watching Emily is a somewhat dystopian experience. One evening, I turned on her stream and she was playing a video game in bed, she then got up and did her hair and make-up in front of the hundreds of people watching. The time difference, she's in Texas, meant that I then went to sleep in London. When I woke up, Emily was still there, now on the treadmill in the gym, promising to go faster and then slower if more people subscribed to her channel. There was something claustrophobic about watching one person go about their restricted life, mostly in a small apartment, even though that existence was being broadcast to the world.

EmilyCC is an extreme example. It would be wrong to couch her experience as an entirely negative one, but she has not exactly picked an easy, or even particularly fulfilling, life for herself. There are other streamers that also push themselves to go harder and for longer. They make money, and sometimes have fun, but at what cost? Perhaps we will only know in years to come, as these relatively young people grow older.

Is this thing on?

When I was writing this chapter, I decided to test out livestreaming myself. One evening, while working on the manuscript, I downloaded some free software that connects to Twitch, hit Go and there I was – available for the world to see. It was both highly disconcerting and somewhat thrilling as people started to find me, which happened almost straight away. I rose to a peak of four concurrent viewers. One person watching my stream started asking me questions and commenting that they could hear my keyboard clacking as I typed away. It was all quite bizarre considering I'd never broadcast on Twitch before, nor told anyone I was there.

The experience was rather addictive too. I hit 'Go Live' when I returned to my desk again the morning after. It's easy to see how people become hooked. You quickly start to think, 'if I stay then some more people will watch', then it's 'maybe I'll get enough followers to be able to start making money on here… but I need to keep going just a little bit longer'. Hours can go by in a blink of an eye and the camera is still on.

Creating videos that you plan, tightly edit and then post on TikTok, Instagram or YouTube leaves the creator with a level of control. Turning on a camera as I typed alone in my flat felt entirely different. I was (potentially) letting anyone with an internet connection into my home, and desperately looking at the viewing figures while I did so. There was a strange sense of empowerment and also vulnerability that came with it.

As you scroll through Twitch, it is hard not to think about *The Truman Show*. The 1998 movie tells the tale of a man (played by Jim Carrey) whose every move is broadcast as reality television. In the film, Truman initially doesn't know what is happening to him. Livestreamers are making an active choice, but it is the authenticity, or at least the pretence thereof, that is of interest to audiences in both Carrey's movie and those watching on Twitch or YouTube.

Zooming through the pandemic

As with so much else, the Covid-19 pandemic was a huge moment for livestreaming, and not just on Twitch and YouTube. With few other ways to get live entertainment, comedians and other entertainers used an array of digital tools to continue performing in front of an audience. Much of this revolved around Zoom. Along with Microsoft's Teams and the messaging service Slack, it became a staple of working from home and then hybrid working arrangements as the pandemic faded.

Zoom enjoyed rapid growth during the early months of the crisis. Announcing its quarterly results in June 2020, the video chat tool revealed it had increased its revenue by 169 per cent year on year and deployed 175,000 new licenses as the pandemic and remote work spread and everyday life, from school classes to funerals, had to be conducted virtually.[16]

Stand-up comedy, devastated as clubs had to lock their doors in March 2020, was one part of the entertainment sector that used livestreaming in a bid to keep going. Comedian Catherine Bohart describes 'a 24-hour period where *everything* went'.[17] Performers grabbed ring lights and webcams and performed from their homes. Working alongside Helen Bauer, Bohart herself launched Gigless, a monthly online comedy night that grew to become popular. 'Having a loyal and excited 100 people every week watching my stuff. That gave me a sense that what I was doing was worthwhile,' she commented to *The Guardian* in February 2022.[18]

Being funny via livestream instead of on a stage was a struggle though. The work doesn't necessarily translate directly from one setting to the other. Comedian Mary Beth Barone said that 'instead of scratching the itch, it was just making it worse'. Brittany Carney said that when performing online 'the timing is all weird'.[19]

Musicians had to take a similar approach to comedians. As club nights, concerts and festivals were cancelled, DJs livestreamed sets, benefit concerts were broadcast online and the likes of John Legend and his wife simply popped up on Instagram Live to chat and play song requests.[20]

Instagram Live actually proved to be the more appealing setting for musicians, perhaps because people could just spontaneously drop into the 'gig' in a way that wasn't possible of Zoom, where a meeting link needs to be set out. Chris Martin played Coldplay hits to viewers. Singer H.E.R launched a series named Girls With Guitars, in which various guests appeared. Never one to hold back,

Elton John put on a huge livestreamed concert in association with iHeartRadio. There was the sense the musicians needed this as much as the fans did.[21]

While these events were mostly better than having nothing at all, not least because many comedians had to keep performing to make any kind of living and Zoom allowed them to do this, it just wasn't the same. As much as livestreaming can connect people, and the reality was that during 2020 and much of 2021 it was the only viable option, sometimes you really all do have to be in the room together and the pandemic showed that.

The dark side of livestreaming

For many, livestreaming is a fun way to connect with people who share similar interests. Yet there is a dark side to it too. One prominent issue is sexism and the misogynistic way women are treated on livestreaming platforms, something noted by a variety of creators in several reports, as well as in academic studies.

There are a multitude of factors at play. Some of it is just an extension of what women generally face online. This is something we have to deal with in the real world too, but it is exacerbated online by the ability to be anonymous on social media platforms. This is then combined with issues of misogyny pertinent specifically to the gaming community, which is at the core of livestreaming. Speaking in 2021, Danielle Barrie, who owned the Dando Twitch account, told *The Scotsman* newspaper: 'One of the main reasons why we see harassment against women in particular on Twitch is because men don't think we should be here.'

Even more sinister, Barrie added that men 'think we are here just for their viewing pleasure and that whatever they ask us to do, we'll do it'.[22]

Racism and abuse of LGBT streamers are similar concerns. Chat moderators, usually just a member/s of a streamer's community, can only do so much. At various points, streamers have called for black-outs of Twitch until the company took dealing with the issues more seriously.[23] Much of this happened in the fraught political climate of 2020 and 2021, but there is no indication that these problems went away in the years after.

Research published in 2025 by Kristel Anciones-Anguita and Mirian Checa-Romero from Universidad de Alcalá in Spain found significantly higher levels of self-sexualization by women on livestreaming platforms compared to men. The researchers say that 'female streamers may resort to self-sexualization as a strategy to engage more followers', because popular categories like gaming are not really seen as being for women.[24]

Twitch also changed its rules December 2023, having loosened its laws on so-called artistic nudity just days before. The new language, published, in January 2024, banned nudity and the suggestion of nudity, but also put a lot of emphasis on correctly labelling streams. There was an exception to the nudity rules if the 'Pools, Hot Tubs and Beaches' label was used. Ultimately, it all actually added to the perception that women could be objectified by others and objectify themselves on these platforms, but could rarely be taken seriously as 'proper' gamers or streamers.[25]

Terror, livestreamed

On 15 March 2019, a man walked into two mosques in the New Zealand city of Christchurch. He opened fire with a semi-automatic weapon, killing 51 people. A further 49 people were left injured.

On 9 October 2021 a man approached a synagogue in the German city of Halle. He was unable to get in, but shot dead a nearby woman and a man in a local kebab shop.

Both these attacks were motivated by racism and hatred, but they share something else in common too. These terrorists broadcast their crimes online. The first was livestreamed over Facebook using a head-mounted camera. The video was framed to look like a first-person shooter video game such as *Call of Duty* or *Counter Strike*. The second attack went out on Twitch. The men behind them also published bile-filled manifestos online. The Christchurch killer's ran to 87 pages. As well as taking lives and causing fear, these attackers were looking for fame and notoriety, which is why I decline to publish their names here.

The Christchurch terrorist flagged to those on the infamous 8Chan message board that he was going to livestream the attack, determined to attract as wide an audience as possible. He was aided by having three mosques situated close together in a city that had strong mobile broadband coverage. This meant he could do maximum damage in a high-quality video. The synagogue attack was live for 35 minutes via an account that had been created just months earlier.

The live streams of the atrocities were just the start. Once they were over, the footage captured could then be spread online. Facebook says that less than 200 people saw the live stream that went out on its platform. Twitch insisted that 'approximately five people saw the German attack there'. Shirley Leitch, analysing the Christchurch massacre in *Rethinking Social Media and Extremism*, wrote:

> The viral success of the Christchurch massacre video produced using cheap, basic technology by one person provides a stark illustration of the challenges faced by social media companies.[26]

Social media companies try to use AI to catch such incidents, but the technology can be bypassed. For instance, because the attack in New Zealand looked like a first-person shooter video game, something that is completely legitimate to stream, the AI struggled to tell the difference between a game and real life. It is fair to assume that such systems will improve over time, but nothing is foolproof.

On top of the tech, human users can flag it up to platforms if they find something violent and abhorrent being livestreamed. Facebook said that none of the people who watched the horror unfolding in Christchurch told them or the police. This is presumably because this audience consisted of supporters of the attacker who had arrived at the stream from 8Chan, although we cannot know that for certain. Twitch said that the video from Germany was flagged after 30 minutes and was then taken down.

Tech companies unequivocally do not want their products used to broadcast terrorism. If nothing else, such incidents make advertisers, who obviously do not want to be associated with such content, cautious about putting ad dollars into such platforms. The industry has proposed ideas such as making the platforms assign a special status to venues such as schools, places of worship and hospitals, with stricter rules for live streams coming from them. That might help, depending on how the location of the video was determined. Another suggestion put forward is that there should be a delay in certain content going live. Whatever the solution, the incidents in Halle and Christchurch revealed the companies' limited ability to actually deal with the problem when it occurs.

Some of this is self-inflicted by the platforms. Major social media companies have significantly reduced the size of their moderation teams, diminishing their ability to have humans spot dangerous content, even those as heinous as livestreamed murder. Many social networks laid off large numbers of staff from these teams, meaning that even if content is flagged by users it will take longer to deal with. In 2023 Free Press, an organization that monitors these issues, flagged the potential risks to the US presidential elections the following year from tech firms reducing this capability.[27] Similar concerns exist around the broadcast of terrorism.

Hashing is another technology used to try to reduce the impact of these horrors. This assigns a digital fingerprint to a video and places it in a database used by all the major social networks. Once one of them hashes a video or image, that content is automatically blocked across the board. This does not affect the initial stream but may stop it spreading after the event. This is important given that the ripple effect is a key motivating factor for broadcasting an attack in the first place, with many more people watching the republished video than the original broadcast.

Online moderation is a thorny issue, one I personally tend to approach from a more liberal, free speech perspective. But there is no debate as to whether the real time broadcasting of racist mass murders should be tolerated by technology companies. Livestreaming platforms are largely used for fun, but they still have work to do in order to deal properly with sexism, racism, homophobia and terrorism.

Making a connection

Despite these terrible incidents and issues, livestreaming provides a lot of joy for a lot of people. It can be fun and bring people from all over the world together.

One particularly lovely moment happened in 2022 when Ryan Stoker, a Brit who streams across Twitch, TikTok and other platforms, encountered a young fellow player while playing *Call of Duty: Warzone* online. The kid admitted he had never got a 'win' in the game before and told Stoker, who he was a big fan of, that he was being bullied at school for not being very good at the game. Stoker had over a million followers at that point, but rejoiced as he helped @extreme_paul get to 100. He told his chat that his stream would not end until the boy scored a first win. He kept finding his new comrade when he was killed in the game, guiding him to that victory.[28] The viral clip of their interaction is really sweet, showing how excited the fan was to meet his hero and claim the win. In a digital world where things can be unpleasant, this was a charming, fun moment that shows how online connections can be powerful. It is unlikely @extreme_paul will ever forget 'meeting' his favourite streamer.

05
Audio battles
or, The beauty of music at our fingertips

Grunts and splutters fill the room as the home computer connects to the internet, the modem making a screeching sound that only readers of a certain age will remember. Clicking on the image of a demonic-looking cat wearing headphones, you realize that almost the entire world of music is now available at your fingertips. Not quite instantly – internet connections are painfully slow at a time when access to broadband was, at best, rare – but it's pretty much all there. No more trips to the record store required. No more hoping that all the tracks on the album you just spent money on are good. No more paying to listen to a band at all, in fact.

Peer-to-peer file sharing service Napster was, for a time in the early noughties, a phenomenon. The free software allowed users to upload their music collection and download music shared by others, although more people were interested in doing the latter than the former. If you wanted to listen again to the track you had just heard on the radio, or check out more of that band's material, you could do so in just a few clicks. The whole thing sent shockwaves through the music industry, which had previously been a comfortable source of profit for those involved.

Napster's legal position was always somewhat dubious. Copyrighted material was being shared without artists or labels being paid, but for a period of time there was very little they could do to stop it. Downloads placed a heavy burden on the internet connections of the day, and the audio quality of the files being shared wasn't always the best, but it was free and so obviously very appealing to users who could master the tech. There was a real thrill in finding the albums you wanted in good quality and them arriving on your computer, however long it took.

Co-founded in 1999 by Sean Parker – played by Justin Timberlake in the film *The Social Network* – and Shawn Fanning, Napster was revolutionary. At the time, the CD business was worth £30.6 billion. The newcomer service launched in the June and gained 150,000 users in just four months. Anything from 26.4 to 80 million users were said to be signed up at Napster's peak in February 2001.

The image of Napster is of a rogue agent, determined to blow up a complacent, arrogant industry. To a large extent, that's true. However, there was a level of professionalism and business acumen within the company from early on that is often overlooked. For example, established venture capitalist Eileen Richardson was the firm's first CEO. She has said that she was planning to launch a model that would have seen users pay $1 per track, something Apple would later make a success of.

Ultimately, Napster could not survive legal challenges from hip-hop superstar Dr. Dre and heavy metal icons

Metallica. It paid £20 million to copyright holders in September 2001 and by June 2002 was filing for bankruptcy protection, having been unable to pay its staff the month before.[1]

While the company itself may not have been able to survive, Napster undoubtedly changed the face of the music industry. Imitators such as LimeWire followed in its wake. Record executives and artists realized that digital was the way forward. What is more, fans had suddenly got a glimpse of the future and many did not want to go back. The question, then, was how to create a digital music landscape that was to everyone's benefit?[2]

Enter Steve Jobs.

Apple rescues the record industry

In his biography of the Apple founder, Walter Isaacson explains that Steve Jobs, a devoted music lover himself, recognized the importance of music as the tech industry was growing. Specifically, he became aware of how much people were burning on to CDs – i.e. taking a digital music file and then putting it on a CD they could play on their usual system. Seeing how much people wanted this functionality led to the invention of first iTunes, then the iPod.[3]

At the time, people were either ripping albums their friends already had instead of buying it themselves or getting the files directly from Napster. The music world was not in a good place. Representatives of Warner and Sony turned up at Jobs's office at Apple HQ to try and present him with some ideas they had. The execs were told, in typically Jobsian fashion: 'You have your heads up your asses.' Somewhat surprisingly, they agreed with him.

Tom Kiehl, chief executive of industry body UK Music, says the sector has 'always been on a journey. I think when you see the evolution from vinyl to cassette to cassette to CD to online, we're always going to go through this kind of moments and disruptions.' He adds that 'you've got to see the debate around music streaming from the very beginning, in terms of the Napster moment around 1999 and the evolution of the debate there'.

Even in the heat of the 'Napster moment', Jobs, of course, had the solution. The Apple co-founder realized that you had to make it easier to get music legally than to pirate it. The premise was simple, achieving it less so, given the huge technological and business shifts that were required. Eventually though, deals were struck with the largest record labels, and the iTunes Store went live on 9 January 2001.

The iTunes Store offered digital versions of individual songs for just $0.99 in the US and £0.79 in the UK. It is hard to overstate the effect this had on the music industry. It led to people embracing digital music and thus to the streaming landscape we see today. It had a big impact on how listeners approach music too. Individual songs had always been available to purchase as singles, but the $0.99 track really broke up the focus on the album as the key unit of consumption.

While Napster had got some people used to downloading music, it was somewhat fiddly to use and not entirely reliable. I recall times when I thought I'd got a song I wanted but upon trying to play it the file turned out to be a dud. The iTunes Store was safe, intuitive and legal. If you were in the Apple ecosystem, using a Mac and an iPad, it had suddenly become easier to pay for the music than not. You also knew you were getting a good quality recording and not running the risk of downloading viruses.

Mission accomplished, then? Well, sort of. For one thing, music piracy still exists to some degree, although it is not the existential threat it seemed to be around the end of the nineties and the early noughties. Kiehl comments that the 'piracy debate's changed':

> We're no longer about peer-to-peer file sharing in quite the same way we were, just because people are not consuming music in that way anymore. There's issues around, say, music stream ripping [which involves using easy-to-access software to download a piece of music being streamed on YouTube or Spotify and keep it permanently], and the use of user generated content, and the way that the kind of piracy impacts around that as well.[4]

The iTunes Store, responding to the initial piracy threat at the end of the nineties, was about offering music as standalone digital purchases, not streaming. It helped get users accustomed to not having physical copies of songs and albums, and getting digital alternatives in an entirely legal manner, but there was no monthly or annual fee to get whatever you wanted. It was just about moving people away from having vinyl, cassettes and CDs. Breaking the concept of ownership entirely would take a few more years.

From iTunes to Apple Music

iTunes lasted until 2024, when, having already been eliminated from Apple's own operating systems and divided into different apps, it was split up in to different pieces of software on Windows too.[5] iTunes had dominated for years though. In its 2008 annual report, Apple spoke of 'significantly increased net sales from the iTunes Store in each of the Company's geographic segments'.[6]

As time went on, Apple offered a growing amount of different content through the software. Things had moved on from being just about music. Videos and apps could be accessed too. Even certain types of device management needed to be conducted through iTunes. This ultimately led to it being a rather tricky piece of software to navigate. After iTunes' demise, Apple even mocked how unwieldy it had ended up becoming during one of its events.

Apple Music was announced on 8 June 2015, and launched at the end of the same month. Eddy Cue, Apple's Senior Vice President of Internet Software and Services, described it as 'all the ways people love enjoying music come together in one app'. As well as being a straightforward music streamer with playlists and other functionality that we expect from such software, Apple Music offers live radio too. It has standalone shows at set times and real, human DJs including no less a figure than Sir Elton John, who has presented hundreds of episodes of his *Rocket Hour* show. The development of Apple's radio offering was pioneered by DJ Zane Lowe, who rose to fame in the UK, largely on BBC Radio 1. He moved to

California in February 2015 to take a lead role in the development of Apple's music streamer.[7]

Lowe's relocation was another demonstration of how the tech titans were moving into the cultural space as streaming grew. He was a hugely popular DJ on the most influential music radio station in the UK with the ability to make or break artists but he left for a tech company. Apple had long had cultural cachet around the music space. It had even had a radio offering before Apple Music. It owned the Beats brand originally launched by Dr. Dre, who had helped the demise of Napster, and offered a live station called Beats 1. But Lowe's arrival and the rebranding as Apple Music took things to a new level. Apple would go on to launch a variety of radio stations focusing on different genres such as country and club music.

In most sectors in which it is involved, Apple is the leader, if not in market share than in cultural dominance, but this is not the case in streaming, where it has long played second fiddle to a much bigger beast.

Spotify

Spotify launched in 2008, although founders Daniel Ek and Martin Lorentzon had been developing it since 2006. The Swedes, like Steve Jobs, recognized the issue of piracy in the music industry and wanted to do something about it.

Ek, who became the face of the company as its long-term CEO, had come to the same conclusion that Jobs had

– that you could only end piracy by making legal access to digital music easier than illegal piracy. 'The only way to solve the problem was to create a service that was better than piracy and at the same time compensates the music industry,' he told *The Telegraph* in 2010.[8] Whether the industry is actually fairly compensated would be a live debate for years, but Ek and Jobs have been proved right in principle.

Spotify was initially available in the UK, France, Spain, Finland, Norway and, of course, Sweden. It arrived in the US in 2011 and expansion accelerated after that. By 2012, there were no restrictions to its US offering and it was in Germany too. This was followed by launches in Portugal, Poland and Italy in 2013 and Canada the year after. Asian and Middle Eastern expansion happened in the following years, with India and its population of over a billion people added in 2019, a particularly significant moment given the size of the market.[9] By 2025, Spotify was in 237 countries and territories, spanning the globe from Åland to Zimbabwe.

Unlike the iTunes Store which it originally looked to challenge, users do not buy and download individual tracks or albums on Spotify. Instead, you pay a monthly subscription to access the music you want. Spotify also maintains an ad-supported tier with reduced functionality. All this might seem totally normal to us now, especially because it is what we do with TV and movies courtesy of the services described in much of the rest of this book; however, when it launched, Spotify marked a profound shift in music consumption.

The growth of Spotify

For users and artists, Spotify might have come to feel omnipresent – an essential subscription to have and a crucial place to make your music available. However, it took a long time for the company to become profitable. In November 2024, the firm published results that showed it was in the green for a full year for the first time, with profit sitting at €1.14 million.[10] It had taken 16 years to get to that point.

The number of Spotify users has continued to climb over the years as it moved toward this point. The service has added more and more features over time too, with podcasts and audiobooks becoming (almost) as important to it as music. You can even watch video through Spotify now. However, like all streaming services, prices have risen as it pushed for that profit. In June 2023, a Spotify Premium subscription cost £/$9.99 per month. A year and two price rises later, it was £/$11.99 per month.

Podcasts and audiobooks

Podcasts have been growing in popularity (and number) for some time. They require an in-depth discussion in and of themselves, something we will do later on. Suffice it to say at this juncture that Spotify realized that this was a space it needed to operate in so that it was not totally dominated by Apple. The first meaningful arrival of non-music content happened in 2015, when podcasts, news bulletins and video clips became available on the app.

In 2020, Spotify showed how seriously it was taking podcasts by splashing the cash on Joe Rogan. The controversial, wildly popular host signed a three-and-a-half-year deal said to be worth $200 million. The agreement made his show exclusively available on Spotify, a huge coup for the streamer. In February 2024, the deal was renewed for $250 million, with the exclusive element removed. At the time, the *Wall Street Journal* reported the deal involved a minimum payment and share of ad sales revenue, with those sales conducted by Spotify.[11]

In September 2023, Spotify said it had '100 million regular podcast listeners, a ten-fold increase since 2019'.[12] At that point, there were 5 million podcasts on the platform. In March 2025, the company claimed to be the home of 6.5 million podcasts.[13] Those numbers show that Spotify's intervention into the podcast market was not just about big names like Rogan.

Podcasts were part of an expansion strategy that was all about making Spotify a one-stop shop, a bit like iTunes before Apple split it into different pieces. With that in mind, it added audiobooks in 2022, a market that was essentially monopolized by Amazon-owned Audible before that. Three years later, there were 350,000 audiobooks on Spotify.[14] These can be bought individually alongside a monthly subscription, giving the company both another lucrative revenue stream and a way to keep users in their app for longer.

Video

As if to emphasize how it has progressed, Spotify has integrated lots of video content too. Video versions of lots of podcasts are available on the service. Similarly, users in a number of non-US countries can watch music videos following the roll-out in March 2024.[15] The company describes video as 'becoming core to the global Spotify experience'. In June 2024 it said that 170 million users had watched a video podcast on the platform[16] and it has even tested video educational courses in the UK.[17] Quite the shift from being 'just' a music streaming service.

Audible

There are various ways to get audiobooks: Apple has a way to purchase them through its Books app alongside the offering from Spotify. However, the undisputed market leader is Audible. The company was founded by Don Katz, who vowed to 'build a new medium that will redefine and enhance the nature of spoken information, education, entertainment, and other modes of verbal expression we will help create ourselves'. Katz, formerly a journalist, recalled on NPR's *How I Built This* that when he was growing his firm in the 1990s he 'kept going to people, who I otherwise respected, who thought I was crazy'. In typical tech entrepreneur fashion, this didn't put him off. Instead, 'it made me more inspired to go ahead and try to do it,' said Katz.[18]

The company that people told him he was crazy to build was sold to Amazon in 2008 for $300 million and audiobooks have continued to grow ever since, at about 30 per cent or more per annum. The market in the UK by some calculations reached a valuation of £1 billion in 2023.[19]

Audible combines a subscription and individual purchase model. Users get one credit per month with a standard subscription and then can bolster their library by buying individual books. Amazon has never folded Audible into Prime. In our hectic modern world, audiobooks are a great way to consume books and are a challenger to the music streamers for our ears.

Other players

Spotify and Apple Music are far and away the most prominent music streaming services, but there are others that have a meaningful user base. It is 'more complex' than just being a plain duopoly, according to UK Music boss Tom Kiehl: 'Often [the popularity of a service] can be on a territorial basis,' he explains.

Separate to Audible, Amazon has its own music service bundled in as part of a Prime subscription and also an Unlimited option, which, in 2025, cost an additional £10.99 per month. Unlike its TV offering, Amazon Music has not really had an impact. Perhaps part of the reason for this is that Amazon music is not a very satisfying experience. The service curates playlists for you based on the artists you have told it you like, which is fine, but unless you upgrade to the Unlimited option you can't simply hit Play on an album you want to listen to. Sometimes you just want to listen to music, not be upsold a more expensive subscription, but that feels like what the entire service is geared towards. Pretty galling if you are already paying for Prime. One of the big advantages that Amazon Music has is that it connects seamlessly with the company's Alexa hardware, although you can connect Spotify and Apple Music to Alexa too.

Deezer initially launched in France then Brazil before expanding into other markets, including the UK and the US. It focuses on strategic partnerships and boasts of being 'the first platform to introduce a new monetization model since the inception of music streaming, designed to better reward the artists, and the music that fans value the most'.[20] It is a perfectly decent offering, with the cost of a subscription set at a similar price to its rivals.

While all of these services have things to recommend them to users, none of them has had the impact that Spotify or Apple Music has. The only alternative that has had any level of mainstream cut-through is YouTube Music. 'You can't kind of take [YouTube] out the equation,' says Kiehl, noting that it also has both free and paid-for options. Research from Kantar released in August 2024 showed that YouTube Music was enjoying significant growth, becoming the 'most adopted music streaming service' in the second quarter of that year. Much of this was driven by incorporating non-music audio such as podcasts and audiobooks. Craig Armer, Global Strategic Insight Director at Kantar, told *Music Business Worldwide*: 'There's a significant opportunity to engage older demographics who are less involved but show considerable potential. The challenge lies in convincing these audiences of the value of paid subscriptions.'[21]

Ultimately, 'there are a number of players in this, platforms and services, which we would consider as fundamental parts of the streaming market,' says Tom Kiehl. Audiences have generally made their mind up though, and they want Spotify and Apple Music.

Streaming in the mix

DJs love to talk about how they can mix CDs or even vinyl records and how that makes them a 'proper' DJ, but in the streaming era it is becoming increasingly unnecessary (unless you really like the sound vinyl offers). As Bill Brewster and Frank Broughton put it in the introduction to their defining work *Last Night a DJ Saved My Life*: 'Aside from the musical knowledge, and the ceaseless research and collecting which supports it, the DJ's skill lies in sharing his music effectively.'

Streaming has proved massively helpful in doing all that. There are a host of services that incorporate a streaming element that are particularly useful to DJs.

SoundCloud is an online service that is a great way for artists, especially up and coming ones, to get their work out. It combines streaming with a social element, as users can follow and even message artists that they like. It offers both ad-supported free options and a paid subscription service. Digging through playlists or DJ mixes, as well as the various artists suggestions that come up in your feed, is a useful way to uncover new music.

SoundCloud users can also link the software with various pieces of popular DJing software. This requires a specific SoundCloud subscription in a couple of instances, but not in all, making in relatively straightforward for a DJ to drop a track they've only just discovered.

Even more specialized is Beatport. Aimed solely at DJs, it combines streaming and purchasing options of tracks in various audio formats. It provides some of the crucial information DJs need to know about a track, such as what key it is in and what tempo it is. Beatport also offers its own online DJ tool, a chat forum and music production software too.

Most people will never encounter or need to encounter these services. A straightforward monthly payment to a mainstream music service will suffice. However, for DJs and others who need to stay on top of the latest developments and play tracks that may not even have been given a widespread or physical release, streaming technology has opened up new possibilities. It also gives artists the opportunity to get their output in front of tastemakers. Having a popular DJ discover your track on SoundCloud or Beatport and then go and play it in a club can make a very big difference to a producer.

Artists being paid

As is the case in other creative industries, the arrival of streaming caused an economic upheaval in music. There are regular complaints from artists that they simply cannot earn enough now. A 2020 poll by the UK's Musicians Union and the Ivors Academy found that 92 per cent generated less than 5 per cent of their income via online streams.[22] (This is part of the reason musicians argue that hugely expensive concert tickets are justified – it's how they make their money.)

Big names in the business have pushed to try and improve things. Sir Paul McCartney, Annie Lennox and Coldplay's Chris Martin were among those who wrote to the UK's then prime minister Boris Johnson in April 2021 highlighting the issue. They argued that 'songwriters earn 50 per cent of radio revenues, but only 15 per cent in streaming'. In this instance, they were calling for a small legislative change to bring streaming income in line with radio income, but the point is a wider one.[23]

Music royalties are a bit of a murky area. Spotify explains that it does not pay per stream. Instead, it takes its net revenue income from Premium subscriptions and advertising minus outgoings such as credit card payment processing and taxes – and then calculates rightsholder's share of what it calls 'streamshare'. It says its finds these by taking 'the total number of streams in a given month and determining what proportion of those streams were people listening to music owned or controlled by a particular rightsholder. Royalties then go directly to the labels who distribute it per their agreement with an artist.'[24] Spotify can legitimately say it does not know what proportion of royalties the artist receives, but the company does know that it paid out $10 billion to the music industry at large in 2024, a tenfold increase over the course of a decade.[25,26]

Apple also uses a streamshare method. It is hard to get exact figures, as there are various things to take into consideration, but most reporting over the years suggests Apple pays artists approximately $0.01 per stream. Whatever the system, Spotify is widely regarded to be the lowest payer of all the streamers. Independent music firm

Duetti found that Amazon was actually the best payer, because Music is bundled in with the rest of Prime. Apple was the next best, paying $6.20 per 1,000 streams. This is thought to be double what artists get from Spotify, was due to the lack of an ad-supported tier and the company generally operating in higher-end markets. Then came YouTube, then Spotify. It's not hard to see where the problem is. The company with 268 million subscribers worldwide is the one that pays artists the least.[27]

Nostalgia

Streaming has proved an incredible way of providing access to more music and enjoying it on the go. However, downloading an album on Apple Music or making a Spotify playlist does not bring with it the tangible element that makes buying a vinyl record or a CD such a pleasure. It has left a significant portion of music fans pining for the good old days.

'I think you've seen growth in things like vinyl, which probably bottomed out around the 1990s,' notes UK Music boss Tom Kiehl. In 2024, 6.7 million vinyl albums were sold, a 9.1 per cent increase from the year before and the 17th consecutive year in which vinyl album sales had risen, which underlines the point Kiehl made.[28] Unsurprisingly, a significant chunk of the vinyl sold that year were Taylor Swift albums. Her 2024 album, *The Tortured Poets Department*, sold 112,000 vinyl copies in the UK, making it the most-sold vinyl album of the century.

'You've seen a big, perhaps to some extent a nostalgia, market with things like Record Store Day and other initiatives around that area where people, particularly big fans of particular artists, will buy into certain content, which may be vinyl-related products, says Kiehl. However, this is an expensive way to publish music.

CDs, meanwhile, are somewhat 'flatlining', he explains. 'There's a big decline from where it was 20 years or so ago, but there still is a physical market for CDs.' Some people even like to hold on to cassettes. Kiehl says that some of the end-of-years statistics suggest that there is a real nostalgia market for CDs, and for cassettes as well.

It is an intriguing dichotomy that cannot simply be explained away by a combination of older generations wanting to hold on to what they know and the hipster tendencies of younger generations. There is a real joy that comes from the hiss of dropping the needle on a record player, a real warmth that comes from the sound profile of a vinyl record as opposed to ultra-compressed digital files. Records, CDs and cassettes give us a physical connection to the artists we love too. Being able to immediately listen to a song that takes your fancy through a small device in your pocket is a privilege not to be taken lightly, but plenty of people do not want to give up some of the other joys that come with fandom, such as building a record collection.

The power of podcasts

Podcasts sit alongside music as a crucial part of many of our digital audio diets. We have our favourite shows, our listening routines and feel like we know the hosts personally. During Covid, podcasts provided a sense of companionship during those solitary walks. In normal times, they provide a bit of peace on the commute and access to insightful voices, information and entertainment. You do not even need to download shows. If you have an internet connection, simply pressing Play starts things, in the same way you don't need to download a show on Netflix to watch it.

There is a very active debate about how the name for such digital radio shows came to be. The first known use of the word 'podcast' is said to have come in a *Guardian* article by Ben Hammersley in February 2004. In a discussion about downloadable digital radio shows Hammersley asks: 'But what to call it? Audioblogging? Podcasting? GuerillaMedia?'[29] However, Dannie Gregoire, a software engineer, also built a piece of audio software called podcaster in 2004.[30]

Reading Hammersley's piece back is like disappearing into a time machine. It rightly highlights how recording, editing and publishing technology became more readily available, allowing people to make shows at a low cost. Hammersley also talks of a newly independent reporter 'combining the intimacy of voice, the interactivity of a weblog, and the convenience and portability of an MP3 download'. He speaks to Jonathan Korzen, then director of public relations for Audible, which at that point was independently pioneering podcasts and audiobooks. Korzan referred to the internet as 'a medium that can garner a great deal of feedback'. That prediction certainly came to pass. Hammersley also talks about software such as RealPlayer, which has long been sent to the digital graveyard.[31]

Serial growth

Writing in 2025, it can sometimes feel like everyone has their own podcast. (I present a couple myself!) To my mind, the key moment that sent podcasts from the strange, quirky things your nerdy friend enjoyed to something that everyone was listening to, or at least talking about, was *Serial*. The show was hosted by Sarah Koenig and the first season followed the conviction of Adnan Syed, which occurred after the murder of his schoolmate Hae Min Lee in 1999.

The first episode of *Serial* premiered in October 2014 and the show enjoyed an initial 12-episode run that expertly tracked the case, never leaving listeners entirely sure as to whether Syed was guilty or not, despite him insisting on his innocence. It returned in 2022 as Syed's conviction was vacated by the Baltimore City State's Attorney's Office and he was released.

The iconic music and cliffhanger line which ended every episode – 'Next time, on *Serial*' – became staples of millennial culture. More importantly, *Serial* showed just what could be done with what was then still a relatively new format.

Since then, countless shows have sprung up, many of which do not follow the investigative reporting style displayed by *Serial* but are instead based around interviews or conversations on key topics between regular hosts. It's almost impossible to know exactly how many podcasts there really are out in the world, but Demandsage, a company that tracks the data, put the number at around 4.42 million in 2025. The number of podcast listeners has grown too. In 2023, 506.9 million people listened to podcasts, reaching 546.7 million in 2024.[32]

The influence of the iPod

The release of the iPod was another key factor in the growth of podcasting (and also part of the naming debate). By 2004, 60 per cent of the entire digital music player music was held by Apple's device, to which MP3 files, including podcasts, could be transferred.

In a somewhat out-of-character move from Apple, it didn't try and lash out for copyright infringement. Instead, in 2005 it released the iTunes Podcast Directory and 'podcast' was declared Word of the Year by the Oxford American Dictionary. (It was 'sudoku' in the UK, perhaps indicating that we on this side of the Atlantic were a bit slower on the uptake.) President George W Bush even started releasing his addresses as podcasts.[33]

The amazing thing about podcasts is that, at their core, the technology behind them is pretty basic. Indeed, the name of the distribution technology, RSS, is an acronym for Really Simple Syndication. Podcast apps, such as Apple Podcasts, might suggest shows they think you will like, but at a fundamental level there is no algorithm pushing podcasts in the way s we see with other digital media. While the iPod may be no more, its audio namesake very much lives on.

Despite their relatively humble origins, podcasts are now big business. The likes of SmartLess Media from Jason Bateman, Sean Hayes and Will Arnett, and Gary Lineker's Goalhanger Podcasts are among those generating huge revenue. Technology journalist Kara Swisher has regularly referred to how much money she has earned through her podcasting endeavours. Shows like *The Daily* from the *New York Times* or Joe Rogan can bring in millions of listeners.

Radio, someone still loves you

You might think that with the number of podcasts and music available on demand, streaming would have killed off live radio. Instead, live radio has made its way into the streaming era in various ways. The most straightforward of these is by stations and media groups launching apps through which they can be streamed. The BBC is perhaps the best proponent of this via BBC Sounds. The app allows listeners to access any of the corporation's many radio stations live, as well as listen back to already-aired shows and BBC podcasts.

Commercial radio groups have also brought their assets together into apps that let listeners stream their stations. UK media company Bauer consolidated assets such as

Absolute, Planet Rock and (my personal favourite) the rock and heavy metal station Kerrang into an app called Rayo. Bauer's rival Global did the same with its Global Player, which offers livestreaming of LBC, Classic FM and Heart, among others.

The pioneer in internet radio is a US company, Sirius Satellite Radio, which became SiriusXM. The company was founded back in 1990, focusing on using satellite technology to deliver radio to cars. It IPO'd in 1993 and has since developed into an internet broadcasting titan.[34] It requires a monthly subscription and is home to a huge number of channels and exclusive shows. These include ESPN Radio, major news outlets like CNN and Fox News and the legendary Howard Stern. The service operates mostly in the US and Canada. It also owns Pandora, an ad-supported audio service.

It's a healthy business. As of 31 March 2025, Sirius XM had approximately 330 million subscribers and around 160 million listeners.[35] Despite seeing a 5 per cent drop in the first quarter of 2025 compared to the year before, SiriusXM's total revenue for that three-month period was still $1.6 billion, generating a profit of $937 million (a 6 per cent drop year-on-year).[36]

A key rival to SiriusXM is TuneIn, which offers a similar kind of product but in more countries (although individual radio stations are subject to geographical locations). TuneIn is also integrated with Apple's Siri technology and the devices that have it installed such as the iPhone, AirPods and HomePod. The deal was announced back in 2019 and was a significant development for TuneIn.[37]

Direct connection

As with livestreaming (perhaps even more so), podcasts create an intimacy. There is a bond between listener and host. That person is right there, directly in your ears, as you travel, walk or do the laundry.

Podcasts are becoming increasingly elaborate though. They are turning to video, largely via YouTube, with hosts in branded, sometimes custom-built, sets. This stops podcasts being something we consume when away from our screen and puts them in direct competition with the video streaming services. These can be great, but there is also a risk that such highly produced shows will lose the intimacy that makes podcasts so special.

06
Streaming around the globe
Are we all watching the same stories?

Much of the mainstream conversation around streaming, including in this book, focuses on the UK and USA. However, the move towards more direct-to-consumer (DTC) models of media has been profoundly impactful in other places around the world too. Many countries have their own services and unique aspects to their streaming ecosystem that are interesting to explore.

India – they don't like cricket, they love it

Uniting a country with a population approaching 1.5 billion people from various ethnicities and religions is, to put it mildly, no easy task. There is, in fact, pretty much only one thing that can bring the whole of India together: the country's obsession with cricket. D&P Advisory wrote in a report ahead of the 2024 Indian Premier League season that 'the Indian Premier League (IPL) is more than just a cricket spectacle; it's a unifying force that captivates India's diverse populace for two exhilarating months each year'.[1]

The action, which sees some of cricket's biggest stars from both India and the rest of the world do battle in the T20 format of the game, is brought to that 'diverse populace' by streaming. A mix of international and domestic broadcasting rivals fought it out for the rights to the competition. In 2022, the rights were bought for a record-breaking five-year deal, with the digital streaming rights costing $6.3 million for each match. At the time, Mihir Shah, Vice President of Media Partners Asia, commented to the BBC that 'the fact that digital rights value was higher than television showcases the scale and future potential of streaming in India'.[2]

The players in India's national cricket team are afforded deity-like status off the field, their every move watched closely by millions. When the team took on its greatest rivals Pakistan in 2025, the match got 600 million views on the JioStar streaming platform, a quite staggering figure.[3]

JioHotstar was formed in December 2024 through a merger of Jio, a streaming property owned by Indian firm Reliance Media, and Hotstar, owned by Disney. James Murdoch, son of controversial media mogul Rupert, was also involved in the project, underlining quite what a big deal it is. Crucially, the creation of JioHotstar meant that the IPL and India's international matches could be streamed on the same platform.

The subscription model used by JioHotstar is an interesting one that differs in some ways from the other sports products investigated in Chapter 3. Jio won the IPL rights as part of the groundbreaking deal in 2022. It then proceeded to let fans watch the competition for free for the first two years. This all changed with the creation of the JioHotstar service, at which point paid subscriptions were introduced.

The cheapest, mobile-only platform cost Rs149 ($1.75) for a three-month plan or Rs499 ($5.82) for the year, reflecting the significantly lower price points expected in the Indian market, as well as the power of mobile there that meant there had to be a mobile-only plan in order to be accessible to as many people as possible. (There are over 1.1 billion mobile connections in India.[4]) Further plans cost anywhere from Rs299 ($3.49) for three months with adverts, to Rs499 ($5.82) for three months without. That last Premium Plan still has advertising during live events, a trait shared with other streamers around the world.[5]

Looking beyond the boundary rope

It might be cricket that grabs the attention of and brings in many of the subscribers, but JioHotstar is also home to a huge amount of other high-end content. Subscribers can access work from Disney, Warner Bros Discovery, HBO, Peacock and Paramount via the service. In May 2025, Sanjog Gupta, JioStar's chief executive for sports, told the *Financial Times*: 'While IPL acts as almost the gravitational pull for the consumer, the idea is once you bring the consumer through the gate, you ensure that the consumer has widened access to this vast content library, enabled by a seamless experience on the platform.'

The plan seems to be working. Within six months of its launch, JioHotstar had nearly 280 million subscribers, meaning it had almost caught up with Netflix globally. There is meaningful competition in India, including from Sony and its SonyLIV streaming service which is home to a number of popular Bollywood movies. JioStar offers Bollywood movies too, but it is going to score a lot of runs thanks to its cricket rights.

China – in a world of its own

While India is a democracy, China is very much not. That means it has an approach to streaming – and media more broadly – that means things are very different from the other markets we have explored so far. Crucially, major international streamers such as Netflix and Disney+ are not available in China.

In June 2024, Wayne Ma, an extremely well-regarded Hong Kong-based journalist for tech site The Information, reported that Apple and Chinese telecoms company China Mobile had held talks about bringing Apple TV+ to China.[6] Apple, and CEO Tim Cook in particular, have extremely good connections in the country (details of which were also revealed by Ma[7]). Nothing has been announced as I pen this, but if anyone is going to find a way in, it is likely to be Apple. President Donald Trump's

2025 trade war, which saw Apple have to move significant amounts of iPhone manufacturing away from China in the face of enormous tariffs, probably did not help the process of bringing its streamer beyond the Great Firewall of China.[8]

Because of censorship and clampdowns on free expression in China, the country is really only serviced by domestic streamers. Among the offerings are Tencent Video, owned by Chinese tech giant Tencent, and iQYI, owned by Baidu, another enormous firm. A similar situation exists when it comes to social video in China, where TikTok doesn't exist but Douyin does. This is despite the fact that TikTok parent company Bytedance is Chinese-owned. Furthermore, the likes of Instagram and YouTube are also banned in China, with equivalents run by domestic companies allowed in their place.

The growth of streaming in China has been boosted by the extensive roll-out of 5G in the country and, as with India, some of the figures associated with streaming in the People's Republic are enormous. According to Ampere Analysis, there were nearly 400 million streaming subscriptions in China at the end of the second quarter 2023. It also found that almost 50 per cent of Chinese households were subscribed to a streamer at that point.[9]

The numbers just keep rising. As of October 2025, Tencent Video was in 110.1 million households, iQYI in 102.73 households and Youku, owned by Alibaba, is another enormous company, in 91.63 households. Following them were Mango TV (69.83 million) and Yangshipin (52.7 million).[10] The latter broadcast the Paris 2024 Olympics. Millions of Chinese sports fans tuned in to watch their nation finish second in the medal table by claiming 40 gold, 27 silver and 24 bronze. Table tennis is the most popular sport in the country, with some 480 million people thought to have watched it during the Games.[11]

State media and censorship

The Chinese Communist Party widely censors content and services from the outside. Searching for information on the Tiananmen Square massacre was infamously banned in China in 2012, for instance, and BBC World News has not been officially allowed in the country since 2021.[12,13] In addition, the administration uses media, including streaming services, to spread propaganda. China Central Television, aptly acronymed CCTV, is China's national news network. The CCTV Video News Agency has a YouTube channel, sharing videos on major news events in China and beyond in ways the ruling party approves of. Although some clips draw decent viewing figures, many of the videos don't muster more than 1,000 views, but they are still there, easily discovered and potentially influential.

Ofcom revoked China Global Television Network's (CGTN) UK broadcasting licence in February 2021. Following an investigation, the regulator Ofcom concluded that it was a state-run institution and that political bodies are not allowed to be in control of UK broadcast licenses. 'We are unable to approve the application to transfer the licence to China Global Television Network Corporation because it is ultimately controlled by the Chinese Communist Party, which is not permitted under

UK broadcasting law,' an Ofcom spokesperson explained at the time.[14] Before this, the outlet was available on Sky and its related streaming services, as well as Freesat, giving millions of Brits access to Chinese propaganda.

The revocation of the UK broadcasting licence was undoubtedly a blow to the Chinese Communist Party, limiting its international broadcast capabilities. While the development of digital services allows those of all political persuasions to make content available, in some circumstances being cut off from the mainstream really does matter.

Into, and out of, Africa

Across Africa, users have access to global services alongside those specific to the continent. Netflix arrived in Nigeria in 2020, beating the likes of Hulu and HBO in becoming the first available in the country.[15] It had already been available in South Africa for a number of years, having launched there in January 2016. Disney+ is also available in South Africa and a host of North African nations such as Morocco and Tunisia, but there are gaps in streaming availability in large swathes of the continent.

It is hardly surprising that there is some disparity in what is available in Africa. The size of the continent, developmental factors and network connectivity are crucial. In some cases, gaps are filled by domestic services.

South African firm Showmax launched a streaming service in 2005. Two decades later, it is available in 44 countries across the continent, from Angola to Zimbabwe.

The platform offers African originals such as *The Mommy Club NBO* from Kenya, *What Will People Say?* from Ghana and *Prince on a Hill* from Nigeria, as well as a host of South African shows. There are plenty of movies available too.

Equally notable is that Showmax make all 380 Premier League football matches in a season available, a very big deal in a football-crazy continent. The service boasts that it is 'the first standalone Premier League mobile streaming service in Africa'. Games are available via one of the more expensive packages, with the prices for all the various Showmax offerings differing depending on the country the viewer is in.

To give some perspective, in a number of countries in which it is available, Showmax costs between $1 and $8 per month. Alongside African content and football, it also offers access to international services including the BBC, ITV and Peacock, meaning it serves as a hub for a vast array of digital broadcasting.

Giving a taste of home

As well as there being streaming services broadcasting to those in Africa, there are streamers that provide access to content from the continent. iROKTOW was founded in London and provides access to both Nigerian and Kenyan movies, showing off the best of Nollywood (Nigerian Hollywood). Some of these are available for free, while others require a paid subscription. It is another important example of streaming being able to step in to fill a gap in the market, in this instance work beloved by millions of people across both the African diaspora and domestic audiences.

Netflix spends a lot of time and money commissioning and creating shows and movies from the various regions it operates in, and Africa is no different. We've had three seasons of reality series *Young, Famous & African*, for instance. Per the streamer, the show follows 'young, affluent and famous A-list media personalities' who come from South Africa, Nigeria and East Africa. The group are an 'aspirational who's who of music, media, fashion and Insta stars'.

Netflix has worked with various filmmakers to tell a variety of stories, including rom-com *Soweto Love Story*, thriller *Heart of the Hunter* and comedy *Kandasamys: The Baby*.[16] As these are Netlix Originals, they are available outside of their native South Africa. There is also a wide range of Nollywood content on the platform.

It all helps bring a little of the continent to the wider world in ways that would simply not have been possible before streaming.

Australia, New Zealand and Canada

Australia, New Zealand and Canada might be separate countries with their own distinct quirks and unique features, but when it comes to streaming, they share a lot of similarities. For one thing, these Commonwealth nations all have strong public service broadcasters, some of which are not all that dissimilar to those in the UK. They all provide freely available streaming services to go alongside their linear output.

Down Under, the Australian Broadcasting Corporation (ABC) has the iview service. It provides access to Australian news, kids' shows and more to people around the world both live and on demand, with some restrictions applying for viewers not in Australia. As with the BBC in the UK, there is political posturing around the funding of the ABC and its associated products. Unlike in the UK, the money that funds the organization comes directly from the government.[17] (The licence fee is an indirect funding model, although the UK government does set the rate of it.)

In New Zealand, TVNZ has the unsurprisingly named TVNZ+. It is more like Channel 4 in that it is free to air but funded commercially via advertising. There are also geo-restrictions, meaning only those in the Land of the Long White Cloud can use the service.

Canada's CBC has Gem, which operates in a similar manner and has a premium subscription service too that users can pay for in order to remove adverts, like ITVX in the UK.

None of this is really a hindrance to those travelling who want to access their usual offering – with a VPN, they can stream a small slice of home.

There is plenty of international content available on these services too. This includes the hit drama *Sherwood* on iview and the legendary British soap *Coronation Street* and comedy *The Marvelous Mrs. Maisel* on CBC Gem. This means it's not too hard for people from all over the world to use the same tech to watch these popular shows, and sometime major movies, for free when they might need a paid-for service at home.

A separate sporting story

Sport is more dispersed across Australia, New Zealand and Canada, with some contests available for free and other events paywalled. All Premier League football matches were available to watch in Australia via Optus when it existed and on Sky Sports in New Zealand. In both instances, these streaming services are linked to linear cable channels. The Champions League is also available on Stan Sport (Australia) and DAZN (New Zealand). In Canada it's all about streaming as Fubo is the place to go for Premier League football and a host of other sport too.

Cricket, hugely popular in Oceania, is thinly spread too. Some services go on Foxtel in Australia, with others on Amazon Prime Video, to name but two. Kayo, owned by Foxtel, is another player in the game. It's a similar story in New Zealand, although not in Canada as specialist streamer Willow TV is the place for cricket fans in the Great White North, making it pretty easy for them to follow along.

Elsewhere, free service 9Now streams Aussie Rules Football, another incredibly popular sport Down Under. This further adds to the collection of services required to access all the content one might want when in the country.

Streaming a shared culture

All in all, streaming means that wherever someone is in the world, they have access to much of the same content as those from countries both near and far. In India, *Friends* fans can head to Central Perk via both Netflix and Prime Video. That same series might happen to be on Netflix in the UK, a number of other European countries and Japan, and Max in the US, but we are all watching the same stories, falling in love with the same characters and laughing at the same punchlines. The catchphrases are understood and repeated around the world, creating some degree of a shared culture.

Equally, non-Western work is now more easily accessible thanks to streaming. There are specialist services for Bollywood, Nollywood and Manga lovers. Some of these movies and series are available via more mainstream apps too, potentially helping these creations find a new audience that previously might not have had access to them.

Netflix is the most popular streaming service around the world, clearing 300 million subscribers in December 2024. Amazon Prime Video is thought to be second with an estimated 200 million subscribers, with Disney+ up to 126 million in March 2025. Tencent Video is the top ranked of the Chinese services with around 117 million subscribers, with iQYI sixth after getting over the 100 million subscriber mark in March 2024. Between them is Max from HBO, with an estimated 113 million subscribers.[18]

With the notable exception of the Chinese streamers, which do not tend to be accessible far beyond Southeast Asia, the most popular services are generally available in the widest range of countries.[19] Consequently, we are closer than we have ever been to having something like global TV.

Close, but not actually there.

Shows and movies periodically move around, sometimes becoming completely unavailable in a territory, which can be deeply frustrating. I experienced this myself when trying to finally catch up with the DC Comics series *Harley Quinn*, only to find it had disappeared from Channel 4 where it had previously been in the UK and the only way to watch it was by buying and downloading it from a provider such as Amazon or Apple. The show was still available to watch in America with a Max subscription, making the whole thing all the more annoying.

The rise of mobile, the spread of the internet and a simple love of the content have made streaming a truly global phenomenon. It gives people around the world shared cultural touchstones to an extent that has never previously existed and access to work they previously might not have come across. Among all the legitimate moans about fragmentation and never-ending subscriptions, this bigger picture should not be forgotten.

07
Who pays the price?

Do customers get what they deserve?

So far, we've looked at the profound shifts in culture, media and sport brought about by streaming, but there are plenty more key questions to explore. Who do these changes benefit? Who loses out? In reality, this all boils down to a single question: who pays?

All over the place

Fragmentation is a key theme in the discussion around streaming. As ever more services have launched, content has been spread increasingly thinly. Quite simply, we all need a far greater number of subscriptions to be able to watch all the things we want to, whether that be epic TV series, blockbuster movies, live sport or even videos on our phones.

In 2023, *Forbes* magazine published research that revealed that, on average, people subscribe to 2.8 video streaming services.[1] Taking that number alongside the subscriber numbers of various services, it's reasonable to assume that in most cases this works out as Netflix plus one or two others. The costs of this can quickly add up, especially as we know that only a few people have totally cut the cord and so continue to pay for traditional cable packages alongside the streamers. Deloitte's 2024 Digital Media Trends report found that consumers may be reaching the limit of how much they are prepared to spend on subscriptions each month.[2] Hardly surprising, given prices keep rising.

Most people can still get most of what they want at a fairly reasonable price, although that is becoming harder and harder to sustain. Household budgets have been squeezed and streamers may find that people consider at least some of them a luxury they can no longer afford.

It can also feel like this is just too much stuff to get through. The sheer number of TV shows and movies available is overwhelming, to the point of actually being off-putting. That feeling you get when you walk into a bookshop or a library and you see all the books you haven't yet read? Now you get that every time you turn on the TV or open your phone and browse through the various TV and audio apps.

This all sounds overly, perhaps unnecessarily, negative. There can be no doubt that we consumers are in many instances getting bang for our buck. Lots of our favourite shows as well as exciting new work are available at the click of a button, but curating all of the available options is no easy task.

The services are full of big hits and hidden gems but because of the amount of work available to view, there is no mono-culture anymore. Certain series can be very, very popular, but there will still be vast swathes of the population who miss them either because they don't have access to the relevant service or because they have overlooked it in among everything else. Even if 'everyone' watches the same show, they do so at a different time. No longer is there just one major drama series that we are all watching on a Saturday evening. The joy of the fractured nature of streaming is that we can all focus on watching only the things that really appeal to us, but it also means that viewership is spread out and there is little that we coalesce around.

Speciality viewing

The lack of a monoculture actually leads us to a key benefit of the growth of streaming technology – the opportunity it provides to create specialist products catering to a small but dedicated fanbase. These products can cover everything from a specific sport to music videos to horror movies. Fans are able to pay to watch the content that they are passionate about but which would be much harder to get hold of in a pre-streaming world. Sometimes these services let subscribers join with other people from around the world who share their interests. This used to be a lot harder in the pre-internet, pre-streaming days.

Take, for instance, Thunderflix. The self-proclaimed 'first heavy metal streaming service' allows fans to watch performances by and documentaries about their favourite bands for £6.66 (per month (of course!). Then there are websites like shudder.com and screambox.com which provide access to a wide range of horror movies. I have already highlighted handball service EHFTV in Chapter 3, an example of this DTC approach in sport.

So, yes, viewers are asked to pay up, but they have more opportunities than ever before to deep dive into their hobbies and interests. This improves access to the things fans love, even if they are not hugely popular outside of a niche community.

Soaring subscription costs

Anyone who subscribes to a streaming service knows that prices are only going in one direction – up. The rises began as streamers battled to turn a profit. It took until around August 2024 for most of the services, or at least the ones we see figures for, to start getting to the breakeven or profitable point. Finally, Paramount+, Disney+, Netflix, Peacock and Max were all able to demonstrate that they were in the green. The same time the year before they had combined losses of $683 million, a position that was clearly unsustainable.[3]

This is all great for the companies and their shareholders, but it adds up for us customers. The cost of Disney+ doubled over the course of the first five years of its existence. It started at $6.99 per month in November

2019 and by late 2024 it was $13.99 per month for an ad-free plan. Apple TV+ went from $4.99 per month to $8.99 in the same time period. There can be no doubt the growing number of services at ever-increasing prices has consumers wondering what they really need.

On top of rising subscription costs, the companies have become much stricter about password sharing. Netflix finally became interested in this, but so were its rivals. For example, in January 2025 Amazon limited the number of devices that one account could be logged in to. It's all an attempt to get as many people as possible to not just watch their programming but to actually take out a subscription and have their own account.

One way some services have worked to reduce churn while increasing their prices is by making themselves the 'home' of something, whether that be a sport, James Bond or, in the case of Paramount+, *Star Trek*. Having strong library content, not just the hottest new thing, is becoming vital in the streaming wars. Viewers can binge a new series and then cancel their subscription. However, if someone wants to watch *Star Trek: Discovery* as their bedtime comfort viewing they need to keep Paramount+ (though some Star Trek series are on Netflix in some places). The companies like to boast about their new shows and movies but the reality is the catalogue content does a lot of the work in attracting and, crucially, retaining subscribers.

Another way the streamers have worked to reduced costs and keep people signed up is by introducing advertising to fund lower subscription prices.

The return of the advertisers

There are multiple models for streaming services, but the options essentially boil down to: totally free; ad supported; a hybrid of advertising and subscription; and totally ad-free with subscriptions. (For simplicity, let's ignore both the UK's BBC iPlayer, which is funded by the licence fee, and that fact that even pure subscriber platforms show advertising during live events, especially sports matches.)

With all these various options available, almost every streamer now incorporates advertising in some way into parts of its standard practice, either doing so at launch or by bringing them in later. Most of those that were originally purely subscription services introduced a whole new tier that is a cheaper option. When Amazon introduced advertising on Prime Video it made that the default, demanding another £2.99 per month to watch ad-free, a move that prompted a widespread backlash.

I emphasize once again that there is nothing wrong with advertising or mixed-revenue models in and of themselves, or indeed ad-based media models more broadly. The problem in streaming is that consumers thought it would be one thing – ad-free and subscription based – but it has morphed into something else, something that looks more and more like traditional TV. People are paying more and more to once again watch in the way they always have. Combined with the fractured nature of the market, the development really does prompt us to consider whether things are getting better for consumers and if we are getting a good deal from streaming. Asking such questions provokes a mixed response.

A survey published by YouGov in March 2025 found that 23 per cent of British adults thought subscription-based video streamers are 'somewhat better' than they were a decade before. At the extremes, 16 per cent thought they were 'much better' and 8 per cent thought they were 'much worse'.[4] While that does not indicate widespread displeasure, it is hardly a ringing endorsement of how things are either. In many ways, people are just going along with the status quo without huge enthusiasm for it.

There is also data that suggests tolerance for ads, so long as introducing commercials helps subscribers save money, i.e. by bringing the cost of a subscription down. In June 2024, Hub Entertainment Research found that 66 per cent of those surveyed would choose an ad-supported option if it saved them $4 or $5, while 34 per cent said they would go for the ad-free option even at a slightly higher price point. The number of those saying they would choose the cheaper option with advertising had risen from 58 per cent in June 2021 and had ticked up every month since then in which the survey was conducted.[5] Those results suggest that as the number of streaming services has increased, people are looking for ways to keep costs down while maintaining their access.

While there are lots of economic factors at play, such as the cost of living crisis, it is noticeable that the move to embrace adverts by viewers correlated with the increase in entertainment options available to them.

Growing ad spend

As more streaming services have embraced advertising, marketers are putting more money into the ecosystem. The streaming advertising market in the UK alone was predicted to pass the $1 billion mark in 2025.[6]

While you might assume that the streamers would want more consumers to sign up to their more costly tiers, it's not quite that straightforward. The top-end brands require a meaningful number of customers to be signed up before they are interested in buying a spot. Consequently, the platforms need there to be lots of users on the cheaper tier to get the kind of ad inventory that generates serious income. They will ultimately be able to earn more per user from those watching ads despite the cheaper initial subscription.

Send them to the kids' club

While much of my focus is on the business of streaming and the cultural impacts it has on adults, children are by no means exempted from the streaming wars. In fact, making output for the very young can prove very lucrative. The clearest demonstration of this is Cocomelon, an unstoppable force in children's entertainment based primarily on YouTube and made by the company Moonbug, which also makes other widely successful kids' shows. (Parents reading this portion are probably shuddering already, the theme tune ringing in their ears.)

For an adult watching it, this programming veers between the psychedelic and the rage-inducing. It is very deliberately designed to keep children engaged via bright colours, tight editing and catchy tunes. There are stories of two-year-olds who are testing out episodes being placed near a 'Distractatron' – a screen showing mundane images of adult life – by researchers working on episodes. Every time they look away from the episode to the adult images, it is recorded so changes can be made.[7] Everything is aimed at keeping children engaged and the YouTube algorithm happy.

So successful has Cocomelon been that Netflix has commissioned multiple seasons of a spin-off series. It tracks the key Cocomelon characters when they're slightly older, becoming toddlers, keeping children in the ecosystem just a little bit longer.

Moonbug Cocomelon also produces Lellobee City Farm and Little Baby Bum, which are similarly popular. We are talking truly astronomical numbers here. During my research, I found an episode of *Cocomelon* that had been up for just a couple of hours and had already generated over 2,100 views. A video that had been up for two weeks had cleared 1.3 million views. Despite this apparent success, there have been multiple rounds of layoffs at Moonbug and not all former employees talk highly of the environment in which they worked.[8]

That show makes the producers money beyond just the YouTube ad revenue. The IP is hugely valuable. There is merchandise galore to be sold as well as tickets to live events. Peppa Pig, another hugely popular children's property, has its own dedicated outlet in London's swanky Battersea Power Station shopping centre and a Peppa Pig World theme park in the UK. These go alongside its nearly 40 million subscribers on YouTube and episodes across Netflix and Paramount+.

YouTube, Netflix and most other streaming services offer a specialized kids' service or similar options that parents can easily set up so that children do not encounter inappropriate material by accident. There is a fierce debate around screen time and its effects on children. Most parents turn to an iPad or similar at some point, but also try to limit the amount of exposure children, particularly very young ones, have to screens. In the US, a variety of systematic issues such as work demands and a lack of access to childcare seemed to have led to non-white parents offering their children more screen time than their white peers. (White children used mobiles devices for 37 minutes a day compared to 100 minutes a day for black children.)

The truth is we may not yet know the full extent of what screen time does to children, as it is still a relatively new phenomenon. Advice from the American Academy of Pediatrics on the topic has been altered too, moving away from calling for an outright ban on screens for children under two.[9] Whatever the official advice, it is almost impossible to keep children away from smartphones, tablets and televisions, and streamers are here to accommodate them and their parents' needs. What, though, will the streaming environment look like as these children become teenagers, and then adults? One fast-emerging technology is going to play a key role.

08
Reacting to a new reality
But what to do about AI?

Artificia intelligence is 'already starting to have a huge impact on the streaming industry, and that influence is only going to grow.' That's what ChatGPT told me anyway. 'AI is going to make streaming smarter, faster, cheaper – and more personal. But it also raises new questions about creativity, authenticity and control,' it concluded. There are plenty of valid concerns about AI and its accuracy failures (so-called hallucinations), but those particular responses are absolutely spot on.

The TV and movie industries have been upended in a number of ways in a fairly short amount of time. As we've seen, the number of services, and the volume of content available on them, has expanded hugely. The entertainment landscape is profoundly different to how it was just a few years ago. This has led to greater amounts of cord-cutting as consumers have realized they no longer need to shell out for expensive capable packages to get (most of) what they want. However, as I pen this book, the biggest shake-up may only just be starting.

AI and its impact

In common usage, AI has become an almost generic term for computers conducting and automating a variety of tasks. For the sake of *Streaming Wars* it is important to have a clear definition of AI. Technology giant IBM describes AI as 'technology that enables computers and machines to simulate human learning, comprehension, problem solving, decision making, creativity and autonomy'. That seems an appropriate starting point for this chapter.

IBM notes that 'applications and devices equipped with AI can see and identify objects'. They also point out that much of the debate, and certainly that which is most relevant to streaming, focuses on generative AI. This is 'a technology that can create original text, images, video and other content'.

AI is powered and complimented by machine learning, complex models that do not just process an instruction from a user but use the information to make predictions or, as IBM calls them, 'interferences'.[1] It should be apparent already quite why all this is having, and will continue to have, profound effects on streaming.

A variety of companies, the likes of OpenAI, Microsoft and Google, have spent vast sums of money battling for AI supremacy. In his letter to shareholders reflecting on 2024 (released in April 2025), Amazon President and CEO Andy Jassy provided a good insight into the speed at which these companies are operating. He wrote that the huge impact of AI across almost every sector 'won't all happen in a year or two, but it won't take ten either. It's moving faster than almost anything technology has ever seen'. This is a bold statement which, given the regularly bland nature of these corporate documents, really stands out. Jassy also revealed that there were more than 1,000 generative AI applications being created across the company he leads. He said that streaming both video and music were areas in which Amazon was 'aiming to meaningfully change customer experiences' through AI.[2]

Alongside the more 'traditional' tech giants, the streaming companies such as Netflix have made major investments in machine learning research. As Jassy's comments outlined, the aim is ostensibly to provide users with a better service. But as those signs during the Hollywood strikes point out, such tools could be used to replace people in the entertainment industry and save an awful lot of money. They do have some value for consumers though.

Evidence of AI's use in higher-end markets emerged fairly quickly too, even while the technology was still in its relatively early stages.[3] *Emilia Perez* and *The Brutalist* were both nominated for Oscars in 2025. Both use technology to change actors' voices. Star Adrien Brody picked up the Best Actor Academy Award for his performance in *The Brutalist*, even though his voice was altered by AI, presumably, and successfully, with the intention of claiming that very prize. Is this cheating or a clever use of technology, in the same way films use special effects and CGI? I'm inclined towards the more generous explanation, but it's an area that is only going to grow murkier as the technology improves. How will we know where the performance ends and the AI alterations begin?

The initial excitement around AI was largely based on the launch of text-based products. Users typed in queries and received written responses. Things move fast though.

On 9 December 2024, OpenAI announced: 'Sora is here'.[4] A version of this text-to-video tool had emerged back in February of the same year, but this was Sora Turbo. The company claimed that this iteration is 'significantly faster' than its predecessor. The tool can generate videos of up to 20 seconds long in a variety of aspect ratios and in High Definition (1080p). Clearly, this is not quite the basis for a tool that major studios will enter a bidding war for. Yet.

Visual aids

Even the early iteration of Sora was capable of creating shorter YouTube and social media content and was powerful enough to freak out people in the audio-visual creative industries. There was a sense that some had been caught off-guard, even though selected visual artists, designers and filmmakers had been used as testers for an earlier version of the production (some had leaked access to it as a form of protest).[5] Ultimately, the streaming companies and those who create for them are going to have to work out which AI tools to ignore, which to fight and which to embrace.

AI is a more complicated issue for YouTube because anyone can post content there. AI can be used to mislead browsing viewers. In November 2023, the company said that it was 'starting to require creators to disclose when they've created altered or synthetic content that is realistic'.[6] Various other online services have taken a similar approach, but there is no doubt that identifying such work if people do not make honest disclosures is going to get more and more complicated.

The use of AI goes beyond creating the shows, movies and YouTube videos that we watch. Netflix uses the technology to maintain the quality of the stream. It deploys AI for adaptive bitrate streaming. This means that in real time it is monitoring a user's internet speed and the capabilities of the device they are watching on to maintain video quality and avoid those frustrating buffering interruptions, which inevitably happen at a key moment in the story.[7]

Play something. Anything

We have all had the experience of settling down on the sofa, turning on the television and, after being greeted by an array of apps filled with an endless amount of programming, feeling totally overwhelmed. In the end, you end up watching nothing and swiping on your phone or old-fashioned channel surfing instead.

Netflix thought AI could help solve this problem. In April 2021, it rolled out a feature called Play Something. A button with those words on it was built into users' Netflix homepage and when it was hit a random TV show or movie would play. Netflix made the selection by bringing together information on what a user had started, begun and not finished and had in their watchlist.[8] In relative terms, this is a fairly rudimentary use of AI. However, Play Something revealed a determination from the company to use all the tools and its disposal to keep viewers locked in.

The launch of the feature came in the midst of the Covid-19 pandemic. Use of streaming services had skyrocketed and people claimed to have 'completed Netflix', implying they had watched everything that was available to them from the service. Play Something was an attempt to counter both that and the slowdown in new subscriber sign-ups.

Away from this specific Netflix feature, content suggestions are at the heart of how streaming started using AI. They do, though, have to tread a fine line. Consumers want help to not be overwhelmed but also like having control and input into what they watch. 'Consumers often derive pleasure from their own decisions,' write academics Ana Rita Gonçalvesa, Diego Costa Pinto, Saleh Shuqair, Marlon Dalmoro and Anna S. Mattila. 'The initial intended benefits of AI from streaming platforms… can backfire and generate consumer reactance,' they add. This happens if the platforms 'weaken the sense of autonomy consumers seek in their decision-making processes'. Viewers want good suggestions, but do not want to feel powerless in picking what to watch.

The researchers also found that the platforms have to be very careful when launching these tools. Viewers are not forgiving if AI suggests content they are not interested in because not only does it build on a sense of lost independence, the technology has not even worked.[9]

Another key issue with AI-powered streaming suggestions is that they can trap you in a bubble. The aim is to keep you watching as much as possible for as long as possible. So if the algorithm learns you like rom coms or true crime documentaries, that is what it will serve up. On YouTube things can get dark quickly, with regular stories about users being served up increasingly radicalizing content. On top of that, if a viewer branches out just a little bit to watch one specific thing, the entire algorithm can change and suggest content you're generally not that interested in, frustrating the user.

There is no doubt that these kinds of suggestion tools are going to become more intelligent and, therefore, more useful. The reality is that we need more sophisticated aides in order to find things we want to watch without becoming blown away by all the different options. However, the companies need to embed this technology without taking away user autonomy and nuance. Or they at least need to convinced us that we have not lost our freedom to choose.

Searching for the answer

Netflix is also building more AI into other parts of its app too. In May 2025 the company announced it was starting to build generative AI into its search functionality. One of the first features tested was allowing the user to engage with the app in a natural, conversational manner by saying something like 'I want to an exciting action movie' and getting the results back. This was initially rolled out as a test version on Apple's iOS operating system for the iPhone, but there is no doubt that it was the first step and more such tools will appear and become more sophisticated in the future.[10]

AI and audio

Audio streaming has been affected by AI in many of the same ways video streaming has. There is an ever-growing amount of AI content, and the streaming services are using the tools to provide better recommendations to users.

At the beginning of 2025, music streamer Deezer revealed that it received 10,000 fully AI-generated tracks every day. The company worked to develop tools that highlighted this to listeners and remove such work from algorithmic recommendations. Announcing this, CEO Alexis Lanternier said that 'with a growing amount of AI content flooding streaming platforms like Deezer, we are proud to have developed a cutting-edge tool that will increase transparency for creators and fans alike'.

It is transparency that is key. There is nothing wrong with using new technology to create art, but consumers do have a right to know what is going on. Upon launch, the company said it was going to develop its technology so that it could flag deepfake voices too.[11]

Last night an AI DJ saved my life

Spotify uses AI to help with recommendations, curating playlists, many of which are based on 'mood'. One of its most high-profile uses of the technology was the launch of its AI DJ in February 2023. The company described it as 'a personalized AI guide that knows you and your music tastes so well that it can choose what to play for you'. As with Netflix's Play Something, Spotify's DJ used AI to provide music to a listener who wasn't quite sure what they wanted coming out of their headphones.

Based on what the technology had learnt a user was into, it picked a stream of tracks – highlighting new music and flagging work the tech knows the user likes already.

It also used generative AI technology from OpenAI to provide information on the artist, music and genre being played.[12] In June 2022 Spotify acquired the AI voice platform Sonantic.[13] This went into the DJ too, turning text into audio.

It's worth noting that Spotify is a long way ahead of rival Apple Music when it comes to incorporating AI into its offering.

The AI backlash and unlikely collaborations

Music, TV and film creators all have concerns about AI, as do authors and news organizations. The AI companies are not particularly transparent about what data they hoover up in order to train the Large Language Models (LLMs) that fuel their products. Given the mass digitization of content, it would not be difficult to use streaming platforms' catalogues to help build tools like Sora or to create audio and video copies of certain actors.

This has a variety of implications. Not long after AI tools started appearing, there were concerns about fake videos going around. This didn't just affect actors; victims of such videos included former US President Barack Obama, as falsified clips of him speaking swept around the web.[14]

In December 2023, the *New York Times* decided to sue OpenAI and Microsoft, alleging copyright infringement. Many in the publishing industry cheered them on. Actor Sarah Silverman was also involved in suits against OpenAI and Meta. In May 2025, the *New York Times* agreed an AI licensing deal with Amazon.[15,16]

Similarly, some parts of Hollywood have decided to embrace AI, or at least accept that its rise is inevitable. Lionsgate, a venerable entertainment brand that has been largely left behind by the rise of streaming, announced in September 2024 that it had signed a deal with applied-AI research company Runway. The deal meant that Runway would create a new customized model built on Lionsgate's catalogue of content, which includes famous franchises such as Rocky, Twilight and The Hunger Games. The model could then be used by filmmakers, directors and others working for the studio.

Announcing the deal, Lionsgate Vice Chair Michael Burns declared his company's new partner 'visionary', and 'best in class'. He insisted that 'several of our filmmakers are already excited about [AI's] potential applications to their pre-production and post-production process'.[17]

It wasn't just Lionsgate getting into bed with AI either. James Cameron joined the board of StabilityAI, a company that has a video creation model, the same month as the Lionsgate-Runway deal was unveiled. The multi-award winning director declared that 'the intersection of generative AI and CGI image creation is the next wave' of technology in filmmaking. That prediction is almost certain to bear out, and the likes of James Cameron will be fine when it does. It is those further down the creative food chain who are concerned. And for good reason.

For many, AI is a severe impediment to their career, if not a total disaster. Streaming created the need for lots and lots of content. That meant lots of job creation. Indeed, in September 2021, with the country slowly trying to move beyond the pandemic, the British high-end TV industry raised concerns that they could not hire enough people, such was the demand for show creation.[18] Alongside this, more platforms meant more chances for people to get their work out.[19]

AI could reverse much of that as it replaces roles in the entertainment industry. A 2025 survey of 300 Hollywood leaders hinted at widespread job reduction. Research suggested that those working as sound engineers, voice actors, concept artists and in visual effects were the most at risk.[20] Most of these people don't tend to earn huge amounts, or have particular secure incomes already, yet they often have expertise built up over many years.

Opportunities to shine

Nobody should be a luddite about AI. There are clearly amazing outcomes that this revolutionary technology will be able to provide, including in entertainment in which a number of processes will be automated and sped up. The final product, as we saw with *The Brutalist*, might well end up being better than it would have been without AI. Such benefits do not necessarily trump the concerns, though. Emma Corrin, star of *The Crown* and *Black Mirror*, told *Elle* magazine in April 2025 that 'I think [AI's] terrible, actually. It terrifies me. The loss of original, organic creativity and [not] having to be in a room with a group of people to create something is terrifying.'[21]

It's an entirely reasonable point. However, another consequence of the AI roll-out might actually be that visibly 'human' work becomes more valuable to audiences, to the benefit of Corrin and other creatives. I confess that this idea became something of a bugbear/glimmer of hope for me long before I started writing this book, but there is some evidence to back it up. AI-generated content, whether that's web pages or books flogged on Amazon, are regularly referred to as 'slop'. As more of that is inevitably released, the work provided by visibly human creators will be more desired by audiences. Subscribers to streaming services will want to know that they are committing time and money into watching actors who have put everything into their performance; that those thespians are bringing to life a script that a team of writers have sweated night and day over.

Yes, streamers and studios will want to use AI to cut costs. But they may find viewers demand that what they watch has a tangible human element to it.

Disruption beyond AI

AI is then clearly a profoundly disruptive technology across all manner of industries, including entertainment, for good and bad. But there is more to our new streaming reality than algorithms. As should have become clear from the previous pages, for all the manifold benefits of streaming, consumers are undoubtedly overwhelmed by the amount of content available, the number of subscriptions required and the amount it all costs.

The companies have also had to adapt, putting increasing resources into streaming as linear audiences collapse with varying degrees of success. Some products have fallen by the wayside in a short amount of time. One service to falter was Lionsgate+. The service originally launched in 2018 as STARZ. On 29 September 2025 it was rebranded as Lionsgate+ in 35 countries, including the UK.

At the time, Chairman and CEO Jeffrey Hirsch insisted that 'operating under Lionsgate+ internationally brings a distinct and differentiated identity in an increasingly crowded international marketplace and builds on the brand equity in the Lionsgate name that our extensive research has proven is strong around the world'. By February 2024, the service had shut down in the UK and a host of other countries. Its parent company went on to sign the AI deal discussed above.

Clearly, a name change alongside a distribution deal with Sky was not enough to make Lionsgate+ sustainable in that 'crowded international marketplace'. Not enough people wanted to spend another £5.99 per month on a service whose catalogue was simply not that enticing. This is the central issue that both consumers and companies are having to navigate. There is simply so much on offer now and people are not going to subscribe to everything.

Filling the gaps

When we think of popular content on streaming, we think of the big hits. The shows that 'everyone' is watching such as *Wednesday* or *Adolescence* on Netflix or *Ted Lasso* on Apple TV+. However, if you look at the most viewed charts on Netflix, it isn't really these types of shows that keep people streaming. It is much more middling, dare I say generic, content that maintains the numbers day in day out. Many of these shows and movies wrack-up millions of views while barely causing a cultural stir. As critic Jackson McHenry wrote in Vulture, 'We've moved past the era of great TV into the era of a lot of pretty okay TV.'[22]

It turns out that plenty of people are quite happy to have such content playing in the background. As AI tools advance, it will become ever more tempting for studios to use it to make shows and movies, particularly the kind that viewers are not paying all that much attention to.

Podcasts on YouTube can also fulfil this kind of role, although the hosts would hope that they offer higher quality output too. Having a video podcast playing in the background while you do the ironing or other mundane tasks works really well. You do not need to be focused on watching every second of what it is on, but you are getting something beyond just the audio.

The rise of binge watching

Streaming has profoundly changed our expectations for the way we watch shows and it seems appropriate to finish analysing the industry by looking at this. While previously we were used to a weekly roll-out of a new season, we now expect to watch the whole thing when we want, and if that means powering through it during a single rainy Sunday then so be it! We've had box sets and video on demand before, but these were almost always of shows that had previously aired. Entire new series had never really been available to view in this manner.

In 2020, Jolanta A Starosta and Bernadetta Izydorczyk conducted a systematic review into the phenomenon of binge watching. The literature available to them at the time highlighted that while viewing in this way – they defined binge watching 'as watching between two and six episodes of a TV show in one sitting' – could be enjoyable, there were also some negative aspects that we all need to be aware of. There is certainly great joy that comes from being immediately able to see how a cliffhanger resolves. Watching multiple episodes at a time can also make you feel more connected to the characters. It is, though, clearly unhealthy to be sitting in a dark room by yourself watching TV for hours on end. 'The research implies that, particularly, the excessive forms of binge watching can involve symptoms of addiction, such as lack of control, negative health and social effects, feeling of guilt, and neglect of duties,' the academics concluded.

The rise of such behaviour occurred between 2011 and 2015, as Netflix launched and streaming began to gain popularity and was 'ultimately to become a normal way of consuming TV series among general audiences', according to the research. Starosta and Izydorczyk also noted that 'Multiple studies imply that people at the age of 18 to 39 are more likely to binge-watch than older people.'[23] None of this is particularly surprising, given that young people were more likely to be among the first to take up subscriptions to streaming services and that those services want to keep you in their app for as long as possible.

These companies optimize their products to make that happen, not least by sometimes dropping an entire series in one go.

That's what used to happen anyway. Over time streamers have moved back from releasing seasons in this way. Apple TV+, for instance, almost always releases new episodes on a weekly basis. We became so accustomed to watching multiple episodes at a time that it can now be somewhat frustrating when we don't get to blast through a season in that way, but Apple and its rivals also want you to have to keep your subscriptions for longer and need to balance the two priorities. It's another example of things going back to the way they were, even as the streaming era takes an increasing hold.

Ultimately though, the companies know that however they present their shows, if they are good we will watch them. Consequently, among all the changes brought about by the streaming wars, it is AI and the job losses it will likely trigger that remains the biggest fear of all in the creative industries.

Which brings us back to those protesting on the streets of Hollywood in May 2023.

Conclusion

We got everything we thought we wanted, but who is it good for?

The Hollywood strikes of 2023 underlined just how disruptive the growth of streaming had been, and the increasing influence of AI was going to be, on the entertainment industry. The deal that brought the industrial action to an end included stipulations around the use of AI and bonuses related to streaming performance. It was agreed that digital versions of background actors could not be used without permission. A $40 million fund to pay actors for future streams of TV and movies that they are in was also created to try in some way to resolve the royalties issue.

The central tensions that led to the longest stoppage in Hollywood history didn't go away when members approved that deal though, not least because it only covered a three-year period. As cord-cutting becomes more widespread, as AI models get more sophisticated and as more ways to access entertainment become available, the conflicts between the actors, writers and the studios will become increasingly severe. The placards might get dusted off once again.[1]

The actors and writers also garner plenty of attention in the trade press and beyond. Their less glamorous but equally essential colleagues who work the cameras, build the sets and manage the sound are also affected by technological progress. Many of those jobs are more susceptible to AI. How they are protected, supported and developed in the new world is going to be another key battle in the streaming wars.

A strike by IATSE, the union representing film and TV crew members, was averted in 2024, but that doesn't mean it won't happen in the future as the issues get more acute.[2] The solution is not for these groups, or indeed consumers, to bury their heads in the sand. AI is going nowhere and provides us all with enormous benefits. The technological challenges go beyond AI though. Streaming has had a profound effect on cinema attendance which was previously the basis of the entire movie economy. The whole financial model has had to be rethought as people prefer to take in their entertainment at home.

To say that we can watch whatever we want whenever we want is true, but it is also a massive simplification of the streaming story. Streaming offers opportunities for niche work to be accessible. It means studios take risks they previously might not have, because they are not confined to the structures of a TV schedule. However, we have also seen how quickly content can be pulled from services and disappear. In the case of the *Batgirl* movie, it was never released at all.

Consumers do have a level of flexibility unlike anything we have ever had before. This might mean catching up with a popular drama series after it has aired or, as I have done, watching football matches live while travelling on London's underground. Beyond that, no longer do we have to purchase expensive TV bundles filled with channels we don't watch. It is easy to pick and choose the services you want, taking out or cancelling subscriptions depending on the programming available at any given time. Everyone is increasingly aware of the number of subscriptions available and the mounting costs associated with them, but with a bit of effort you can design your own bespoke package in a way that has simply never been possible before. There really is something for everyone now.

Back to the future

At the time of writing, while there have been some notable deals there has not been a major rebundling in which a host of streaming services come together and offer a single subscription. In a saturated market, companies are competing more fiercely than they ever have before. Prices have gone up, but those market pressures mean the firms have to stay sensitive to what people are willing to pay. There is always a level above which consumers will not go.

Each service wants to have the best content possible, as that is the key way to stand out against competitors. This is obviously another benefit to the viewer.

There is, though, ultimately a sense that some kind of rebundling is how things will end. That at least some of the streamers are going to have to work together to placate consumer frustration over cost and fragmentation. In some cases, this will be easy as various services are owned by the same parent company. In other instances, rivals will have to become, if not friends, then, associates. The closest things came to a major streaming bundle was when three major sports providers planned to work together on a project called Venu. That may have fallen apart in the face of legal pressure from rival Fubo, with ESPN launching its already planned DTC product too. However, there is no doubt other deals will be done, will have to be done, in the future. And the products will actually have to launch.

It is inevitable that more sport is going to be heading for streaming services, in some cases exclusively so. To watch, we will have to sign up to those services in the same way we buy sports TV packages. Combine that with the rebundling mentioned and things may start to look a lot like cable TV again. Not that it is even clear whether or not traditional linear cable TV is itself a viable media or long-term proposition. As the younger generations get older, they simply will not be used to purchasing such products. No wonder advertising money is heading to the streamers.

Culture shift

Streaming and its social media siblings are both part of, and a cause of, an always on, on-demand culture. We work out listening to music, travel accompanied by podcasts, scroll while standing in queues and constantly have things playing in the background. It leaves us entertained, informed… and overstimulated.

The first academic study into the effects of online video services was by Rajibul Hasan, Ashish Kumar Jha and Yi Liu from the Rennes School of Business. It was published by the *Computers in Human Behavior* journal in 2017 and found, among other things, that the recommendation systems used by streamers were a driver of excessive use. Streamers are probably not the worst offenders though. 'While people do use video platforms for pastime and entertainment, other internet enabled services like social media, gaming, chats have a much higher entertainment and pastime capability and hence users with such motives are more inclined to be addicted to these platforms,' the academics noted.[3]

Social algorithms and online video service recommendation systems have all been supercharged in the years since that research took place and AI will continue to make them even more powerful. A lack of self-esteem and self-control were also important factors in excessive usage, showing that the issues go beyond simple technological developments.

Interestingly, the academics, who restricted their study to Amazon Prime Video, Netflix and YouTube, found that, while seeking information was a key driver of excessive use, the desire for a pastime or entertainment was not. It seems unlikely that this finding has held up in the years since, as the pandemic made streaming an essential pastime and source of entertainment.

Smartphones, of course, are central to all of this, having become the hub of our personal and professional lives as we jump between apps and different pieces of content. They are how we both create and consume. We no longer have to be bored, but have to work a lot harder to have some peace. (I recommend stepping away from the screens once in a while and reading a good book. Obviously.)

A new relationship

Streaming has fed our new relationship with art and its creators. No more racing down to a shop, splashing out on your favourite band's new release and heading home to listen to it, fingers crossed that it is good. Now the album just appears on Spotify or Apple Music. If you're not into it, that's a shame, but it hasn't cost you any extra money. We do not own these things – whether it's music, movies or TV series – anymore. Our relationship with them is transient. Arguably, this means we value them less than we once did. However, it also means we can be more experimental with what we watch or listen to. You can start a series or a movie and just stop watching it if you find yourself not enjoying it. Consequently, it is a far smaller commitment and you're more likely to take the risk that you would have if you had to spend money on something you're unsure about.

As we buy far fewer DVDs and records, we do not use these things as a way of telling a story about ourselves in the way we once did. It seems like our Instagram stories and TikTok videos have taken up that function instead.

Another part of the changing culture is the atrophying of casual cinema-going. Big movies are heading to streaming services much quicker than they were before the Covid-19 pandemic. Prior to this crisis, a typical exclusive theatrical window would be somewhere between 75 and 90 days. By 2022, this had dropped to 45 days.[4] Audiences are growing increasingly accustomed to waiting for movies to become available in this manner and are happy to wait a month or two in order to settle down at home and watch.

I did this myself with the mega-hit musical *Wicked*, originally released in cinemas in November 2024. Not fancying two and a half hours surrounded by fidgeting strangers, I waited until the movie became available to stream on Now and Apple TV+ in January 2025 (it had already been made available to purchase). In the US the movie made its full streaming premiere in March 2025, becoming available to watch via US service Peacock at no extra cost a mere four months after the theatrical release.

The cinema industry pushed for a baseline 45-day window in which films would only be in cinemas. Michael O'Leary, who leads trade group Cinema United, told the 2025 CinemaCon convention that 'a clear, consistent starting point is necessary to affirm our collective commitment to theatrical exclusivity'. He argued that there must be a baseline.[5] It is pretty understandable why someone in O'Leary's position would advocate for such a thing, but even as he was stood on stage it felt like the genie was long out of the bottle.

The winners

Writing this in July 2025, it seems clear that streaming of all kinds is swamping linear television and that it is set to continue. But this does not need to look like a digitized, fractured form of cable. What we need is a return to the spirit of innovation, the determination that saw Netflix plough on after being laughed out the room by Blockbuster. Simply mimicking what has gone before would be a missed opportunity.

The creator economy, powered by YouTube alongside apps such as TikTok and Instagram, has shown that innovation at various points, whether that is the growth of intimate vlogs or MrBeast's multi-million dollar productions. Elsewhere, livestreamers welcome people into their world for hours, days or even years at a time. However, even the creator space risks becoming repetitive, as creators copy their peers and TV stars transport the fundamentals of the old world into the new.

It would not do to close a book on such a vibrant space downbeat. Streaming has made creative careers more achievable than ever before as there are so many different ways to get your work out. It has made entertainment more accessible than it has ever been. Streaming companies have shown themselves to be more willing to take risks than their linear predecessors were or could be. Video on demand has broken norms and archaic rules and it will likely continue to do so a long way into the future.

With that in mind, the ultimate winner of the streaming wars is the consumer. It is us.

My favourite TV shows to stream

Over the course of this book I've explored the streaming industry and its impacts. One thing is clear above all else – there is loads of great stuff to watch! Choosing what to settle down to next can be a bit overwhelming, knowing what service it is on even more so. To help, I thought I'd share some of my favourite shows and movies, plus those that I think are essential to watch.

- *Ted Lasso* **(Apple TV+)** The series that made the streamer. *Ted Lasso* tells the story of an American football coach who finds himself coaching a Premier League football (soccer) team. Starring Jason Sudeikis as the title character, it is the defining series on Apple TV+.

- *Mythic Quest* **(Apple TV+)** A comedy set in a video game company, but you don't have to care about video games to enjoy it. Full of quirky characters and brilliant one-liners alongside a few powerful tugs on the emotional heartstrings. It dealt with Covid better than almost any other show.

- *Slow Horses* **(Apple TV+)** A fantastic drama series based on the Slough House series of books by author Mick Herron. It is full of dark twists and some very funny moments too. Gary Oldman stars as the leader of a crew of hapless MI5 agents.

- *She-Hulk: Attorney at Law* **(Disney+)** A rather different addition to the seemingly infinite Marvel Cinematic Universe, with the focus on a feisty female superhero. Sadly only lasted for one season.

- ***Soul* (Disney+)** A beautiful animated movie that went straight to streaming during the depths of the pandemic, it revolves around a middle school band teacher Joe Gardner. He loves jazz but has never quite made it. Things go awry when Joe finds himself… dead. Features an amazing voice cast made up of Jamie Foxx, Tina Fey, Graham Norton and more.
- ***Orange is the New Black* (Netflix)** Arguably the series that launched the streaming era, it tells the story of various characters from within a women's prison. The multi-award winning show raised a huge number of issues, from the treatment of incarcerated women to trans rights.
- ***The Queen's Gambit* (Netflix)** A punchy, engaging mini-series about an orphan who turns out to be a genius chess player. Set during the Cold War and starring Anna-Taylor Joy as the central character, the series is based on the novel of the same name by Walter Tevis. It prompted a surge in interest in chess when it was released in 2020.
- ***Wednesday* (Netflix)** This series is focused on Wednesday, the difficult daughter of the Addams family who is sent to boarding school. It is dark, scary and totally engrossing.
- ***Matt Rife: Lucid* (Netflix)** Rife shot to fame via TikTok and *Lucid* was his first special. The first joke caused plenty of controversy, but much of the set is hilarious. The whole thing is worth watching just for the final line.

- ***Mr McMahon* (Netflix)** A documentary series telling the story of Vince McMahon and the rise of the WWE. You don't need to be interested in pro-wrestling to find this brilliantly put-together docu-series compelling viewing.
- ***Reacher* (Prime Video)** A straight-up action series starring Alan Ritchson as the title character, it was something of a breakout hit for Prime Video. Another adaptation based on novels, this time from author Lee Childs.

My favourite podcasts to listen to

I am, quite simply, obsessed with podcasts. I have them on the whole time. My favourites accompany me when I am walking, travelling, working out and, occasionally, when I'm in the bath too. The connections we have with our favourite hosts and shows are quite something. Here, then, are some of the shows I love to listen to and which I think are crucial to add to your podcast app.

- **Serial** Surely there is nobody reading this book who has not listened to this original, groundbreaking series. If that's you, please go back and listen to it… now! It's always worth dipping back into to remind yourself how it influenced the industry.

- **The Bill Simmons Podcast** Simmons has been a leading voice in American sports broadcasting for decades, and his twice-weekly show is an essential listen for fans of the NFL, basketball and more. There is also plenty of pop culture and comedy thrown in for good measure.

- **On With Kara Swisher** Kara Swisher is one of my journalistic heroes so I devour her shows. With *On*, she speaks to some of the biggest names is technology, culture and politics, pressing them hard on key issues in her inimitable style.

- **The View From the Lane** This twice-weekly podcast from *The Athletic* focuses on the football team I support, Tottenham Hotspur. It's my go-to in the run-up to, and the aftermath of, matches. (The outlet makes shows for lots of teams, so pick your favourite!)

- **Mixed Signals** Hosted by Ben Smith and Max Tani, this weekly show covers all aspects of the media. The dynamic between the hosts, two of the best media reporters around, is fun, and their impressive Rolodexes means they have great guests.
- **Kill List** One of the best investigative limited series I've ever listened to. It tells the story of how a journalist, Carl Miller, discovered a list of people that others wanted to be murdered, and what he and his colleagues did about it. If you loved *Serial* and *Missing Cryptoqueen*, this is an essential listen.
- **Land of the Giants** Season two of this show, in which Peter Kafka and Rani Molla tell the story of Netflix, was an inspiration for parts of this book. Other series that focus on companies such as Apple and Disney are also worth a listen.
- **The Grill Room** Puck's media show is a must-listen for anyone who covers, is involved in or simply just intrigued by the industry. Host Dylan Byers speaks to some of the biggest players. Sister show *The Varsity*, hosted by John Ourand, covers sport and is also excellent.
- **WSJ Tech News Briefing** Sometimes, you just want to quickly know what is going on, and this podcast astutely sums up the biggest stories in the tech industry in under 15 minutes every single weekday. Featuring some of the world's best tech reporters, it breaks things down really well.
- **RHLSTP** Richard Herring was one of the first stand-ups to properly get into the podcasting game, and his

> show has been going from strength to strength ever since. He has top comedic talent on as guests and the shows are recorded in front of a live audience, making it feel like you've gone to a gig.

These are some of my favourite shows and podcasts, but I'd love to hear about yours too – I'm sure there are some great things that I've missed! Why not drop me a line – I have the username **charlotteahenry** across most popular social media platforms. Perhaps you'll introduce me to my new favourite watch or listen.

NOTES

Introduction

1 J K Murphy. The best picket signs from the 2023 Writers Strike, *Variety*, 16 May 2023, https://variety.com/lists/wga-writers-best-picket-signs-2023-strike/ai-more-like-ay-yi-yi/ (archived at https://perma.cc/VWZ3-FP55)
2 J Turner. Trimmed trees outside LA studio become flashpoint for striking Hollywood writers and actors, AP, 20 July 2023, https://apnews.com/article/hollywood-strikes-trees-a590deef009171b39a8f6001d63cae70 (archived at https://perma.cc/47MM-6G6Q)
3 D Sunnebo. British streaming market shows signs of recovery after a turbulent year, Kantar, 30 January 2023, www.kantar.com/inspiration/technology/british-streaming-market-shows-signs-of-recovery-after-a-turbulent-year (archived at https://perma.cc/GAU2-69HG)
4 K Buchanan. Netflix buys Kevin Spacey show, rewrites TV rules, Vulture, 15 March 2011, www.vulture.com/2011/03/how_netflix_buying_kevin_space.html (archived at https://perma.cc/8886-TRC9)
5 J Alexander. The entire world is streaming more than ever – and it's straining the internet, The Verge, 27 March 2020, www.theverge.com/2020/3/27/21195358/streaming-netflix-disney-hbo-now-youtube-twitch-amazon-prime-video-coronavirus-broadband-network (archived at https://perma.cc/EB9L-GVKF)
6 J Valinsky. Leo Messi sparks a surge in Major League Soccer subscription sign-ups, CNN Business, 5 September 2023

https://edition.cnn.com/2023/09/05/business/leo-messi-apple-tv-signups (archived at https://perma.cc/RVJ3-Z23Q)

Chapter 1

1 A Weprin. Netflix Has Mailed Its Last DVD, *Hollywood Reporter*, 29 September 2023 www.hollywoodreporter.com/business/digital/netflix-mails-last-dvd-shuts-down-1235603936 (archived at https://perma.cc/2SJ6-EAYR)
2 M Randolph. LinkedIn, March 2023, www.linkedin.com/in/marcrandolph/recent-activity/all (archived at https://perma.cc/RPC3-GNYQ)
3 About us, Blockbuster, https://bendblockbuster.com/about (archived at https://perma.cc/W6G9-8FSN)
4 K Turnquist. The last Blockbuster video store in Bend inspires a Netflix show, nostalgia-loving customers, and more, Oregon Live, 22 February 2022, www.oregonlive.com/entertainment/2022/12/the-last-blockbuster-video-store-in-bend-inspires-a-netflix-show-nostalgia-loving-customers-and-more.html (archived at https://perma.cc/RV8F-SAY9)
5 F Schruers. Steven Van Zandt stars in Netflix series "Lilyhammer", Reuters, 6 February 2012 www.reuters.com/article/lifestyle/steven-van-zandt-stars-in-netflix-series-lilyhammer-idUSTRE815249 (archived at https://perma.cc/A7X3-ND7X)
6 'Sharing the Streaming Road' – Leichtman Research Group, Q3 2019
7 N Sherman and J Clayton. Netflix loses almost a million subscribers, BBC News, 20 July 2022, www.bbc.co.uk/news/business-62226912 (archived at https://perma.cc/J5AC-4464)
8 27% of streaming video services are shared, 20 March 2023, LRG, www.leichtmanresearch.com/wp-content/uploads/2023/03/LRG-Press-Release-03-20-2023-1.pdf (archived at https://perma.cc/JU5C-7UC2)

9 A Sherman. Netflix estimates 100 million households are sharing passwords and suggests a global crackdown is coming, CNBC, 19 April 2022, www.cnbc.com/2022/04/19/netflix-warns-password-sharing-crackdown-is-coming.html (archived at https://perma.cc/XV9F-RS2N)

10 Our newest plan – now available from $6.99 a month, 3 November 2022, Netflix, https://about.netflix.com/en/news/our-newest-plan-now-available-us (archived at https://perma.cc/QA5F-D92E)

11 T Spangler. Netflix's new ad tier looks to be off to a tepid start, Variety, 22 December 2022, https://variety.com/2022/digital/news/netflix-with-ads-subscriber-signup-research-data-1235468044 (archived at https://perma.cc/CX99-QFV2)

12 L Shaw. Netflix's advertising challenge: It isn't big enough, Bloomberg UK, 28 June 2024, www.bloomberg.com/news/newsletters/2024-06-23/netflix-s-advertising-challenge-it-isn-t-big-enough?sref=0BM9MCxt (archived at https://perma.cc/KCY3-BRKW)

13 Netflix. 14 May 2025 https://about.netflix.com/en/news/netflix-upfront-2025-the-center-of-attention (archived at https://perma.cc/H4PS-UHX2)

14 B Steinberg. Netflix says ad tier reaches 70 million users globally, Variety, 12 November 2024, https://variety.com/2024/tv/news/netflix-ad-tier-reaches-70-million-global-users-1236207015 (archived at https://perma.cc/D785-95VP)

15 T Maglio. Ad-free Netflix users watch 40 percent more than ad-supported subscribers, IndieWire, 18 September 2024, www.indiewire.com/news/analysis/ad-free-netflix-watch-time-study-1235048783 (archived at https://perma.cc/ZYF8-NHB6)

16 T Harrington. Netflix to carry TF1: Now an aggregator, Enders Analysis, 19 June 2025, www.endersanalysis.com/

reports/netflix-carry-tf1-now-aggregator (archived at https://perma.cc/4GCQ-BUFJ)
17 How Netflix's recommendations system works, Netflix Help Center, https://help.netflix.com/en/node/100639 (archived at https://perma.cc/JF3N-JPB3)
18 G Smith. Netflix Won't Share Crucial Data, So TV Producers Are Piecing It Together Themselves, Bloomberg, 6 May 2020 www.bloomberg.com/news/articles/2020-05-06/how-many-people-watch-netflix-shows-tv-producers-don-t-know?sref=0BM9MCxt (archived at https://perma.cc/2X7M-XLH4)
19 N McAlone. Netflix is so secretive even its directors don't know how many people are watching their shows, Business Insider, 24 November 2015, https://finance.yahoo.com/news/netflix-secretive-even-directors-dont-203459944.html (archived at https://perma.cc/FHW8-LPF7)
20 T Gerkens. Netflix apologises as Love is Blind reunion show delayed, BBC, 17 April 2023 www.bbc.co.uk/news/technology-65298352 (archived at https://perma.cc/CLZ6-PYYL)
21 W Lee. Netflix's Reed Hastings steps aside as co-CEO, stays on as executive chairman, *Los Angeles Times*, 19 January 2023, www.latimes.com/entertainment-arts/business/story/2023-01-19/netflix-fourth-quarter-earnings-reed-hastings (archived at https://perma.cc/A7FN-7HU4)
22 Associated Press. Netflix experiences streaming delays leading up to Tyson-Paul fight, ESPN, 16 November 2024 www.espn.co.uk/boxing/story/_/id/42418997/netflix-experiences-streaming-delays-leading-tyson-paul-fight (archived at https://perma.cc/C8FU-69M2)
22 S Battaglio. Will Netflix get into the TV news business? Here are the pros and cons, *Los Angeles Times*, 6 June

2024, www.latimes.com/entertainment-arts/business/story/
2024-06-06/will-netflix-get-into-the-tv-news-business-here-
are-the-pros-and-cons (archived at https://perma.cc/
Q32T-HWUG)
23 Ibid.
24 R Keegan. Why a Netflix special is just the start for many stand-ups, *Hollywood Reporter*, 25 June 2019, www.hollywoodreporter.com/news/general-news/why-a-netflix-special-is-just-start-stand-ups-1219342/ (archived at https://perma.cc/JV38-DU79)
25 Ibid.
26 E Meyer and R Hastings. *No Rules Rules: Netflix and the culture of reinvention*, Penguin, 2020
27 www.espn.co.uk/boxing/story/_/id/42418997/netflix-experiences-streaming-delays-leading-tyson-paul-fight (archived at https://perma.cc/C8FU-69M2)
28 J Stebbins. Netflix is expanding its push into video games, but few subscribers are playing along, CNBC, 8 August 2022, www.cnbc.com/2022/08/06/netflixs-video-game-push-sees-few-subscribers-playing-along.html (archived at https://perma.cc/M5ZU-X2VD)
29 J Goldsmith. Netflix Making Hay With Games Based On Its IP, Will Release One A Month Starting In July, Deadline, 18 July 2024 https://deadline.com/2024/07/netflix-video-games-ip-based-emily-in-paris-squid-games-1236014986 (archived at https://perma.cc/7TVP-MEUR)
30 Netflix Q3 2022 Shareholder Letter, 18 October 2022, https://s22.q4cdn.com/959853165/files/doc_financials/2022/q3/FINAL-Q3-22-Shareholder-Letter.pdf (archived at https://perma.cc/6X3R-A2NA)

Chapter 2

1. C Chen. Amazon, 18 October 2023, www.aboutamazon.com/news/workplace/first-amazon-office-jeff-bezos-garage (archived at https://perma.cc/TN8W-LBLS)
2. J Del Rey. The making of Amazon Prime, the internet's most successful and devastating membership program, Vox, 3 May 2019, www.vox.com/recode/2019/5/3/18511544/amazon-prime-oral-history-jeff-bezos-one-day-shipping (archived at https://perma.cc/26JM-C3DB)
3. C Chen. Amazon, 18 October 2023, www.vox.com/recode/2019/5/3/18511544/amazon-prime-oral-history-jeff-bezos-one-day-shipping (archived at https://perma.cc/26JM-C3DB)
4. Amazon and MGM have signed an agreement for Amazon to acquire MGM, Amazon Press Center, 26 May 2021, https://press.aboutamazon.com/2021/5/amazon-and-mgm-have-signed-an-agreement-for-amazon-to-acquire-mgm (archived at https://perma.cc/MY84-BP9U)
5. Mergers: Commission approves acquisition of MGM by Amazon, European Commission Press Release, 5 March 2022, https://ec.europa.eu/commission/presscorner/detail/%20en/ip_22_1762 (archived at https://perma.cc/6JFA-SURF)
6. T Haselton. Oprah Winfrey had the best explanation for what the weird Apple TV event was really about, CNBC, 26 March 2019 www.cnbc.com/2019/03/26/oprah-had-best-explanation-for-what-apples-tv-event-was-really-about.html (archived at https://perma.cc/R6NL-A5LN)
7. Jamie Erlicht and Zack Van Amburg joining Apple to lead video programming, Apple Newsroom, 16 June 2017, www.apple.com/uk/newsroom/2017/06/jamie-erlicht-and-zack-van-amburg-joining-apple-to-lead-video-programming/ (archived at https://perma.cc/JVJ6-XZ7H)

8 Apple's "CODA" wins historic Oscar for Best Picture at the Academy Awards, Apple, 27 March 2022 www.apple.com/uk/newsroom/2022/03/apples-coda-wins-historic-oscar-for-best-picture-at-the-academy-awards (archived at https://perma.cc/YNB5-NE58)

9 K Scanlon. What it takes to get paid by YouTube, TikTok and other social platforms, Digiday, 28 February 2025, https://digiday.com/marketing/what-it-takes-to-get-paid-by-youtube-tiktok-and-other-social-platforms/ (archived at https://perma.cc/2944-7L4Z)

10 G Malinsky. 57% of Gen Zers want to be influencers – but 'it's constant, Monday through Sunday,' says creator, CNBC, 14 September 2024, www.cnbc.com/2024/09/14/more-than-half-of-gen-z-want-to-be-influencers-but-its-constant.html (archived at https://perma.cc/JHQ2-U9GY)

11 Chris Stokel-Walker. *YouTubers: How YouTube shook up TV and created a new generation of stars*, Canbury Press, 2019

12 T Spangler. MrBeast in talks for TV show on Amazon's Prime Video, Variety, 22 January 2024, https://variety.com/2024/digital/news/mrbeast-reality-show-amazon-prime-video-1235882230/ (archived at https://perma.cc/52K3-47JV)

13 M Malone Kircher. Willing to die for MrBeast (and $5 million), *New York Times*, 2 August 2024, www.nytimes.com/2024/08/02/style/mrbeast-beast-games-competition-show.html (archived at https://perma.cc/FN8F-34QV)

14 D Frankel. Streaming Wars settled for now, with Netflix on top – analyst, StreamTV Insider, 13 February 2025, www.streamtvinsider.com/video/streaming-wars-settled-now-netflix-top-analyst (archived at https://perma.cc/FG9S-W8CP)

15 L Elber and M Kennedy. NBC cancels Megyn Kelly's show after blackface controversy, 26 October 2018 https://

apnews.com/article/a84a7250b109411591ed6b976be800a0 (archived at https://perma.cc/822S-KQT9)
16 A Sherwin. Piers Morgan: Why I turned down Rupert Murdoch's generous job offer, *i Paper*, 26 January 2025, https://inews.co.uk/news/piers-morgan-turned-down-rupert-murdoch-job-offer-3493346 (archived at https://perma.cc/35W6-62E9)
17 T Spangler. Amazon's Prime Video Is Adding Apple TV+ as a Subscription Add-On, Variety, 9 October 2024 https://variety.com/2024/tv/news/amazon-prime-video-apple-tv-plus-1236173900 (archived at https://perma.cc/KGB2-J3CN)
18 How to watch Apple TV+ on Prime Video, Amazon, 5 December 2024 https://www.aboutamazon.co.uk/news/entertainment/apple-tv-+-prime-video (archived at https://perma.cc/6P2P-9DKL)
19 United States Securities and Exchange Commission, Form 8-K, Current Report, Pursuant to Section 13 or 15(d) of The Securities Exchange Act of 1934, 10 September 2019, www.sec.gov/ix?doc=/Archives/edgar/data/320193/000032019319000093/a8-kseptember201991019.htm (archived at https://perma.cc/XT3H-3CL2)
20 www.businessinsider.com/disney-ceo-explains-why-he-left-apples-board-cnbc-report-2019-9 (archived at https://perma.cc/M88M-GWVN)
21 Disney and ITV announce first-of-its-kind strategic relationship. ITV Press Centre, 10 July 2025, www.itv.com/presscentre/media-releases/disney-and-itv-announce-first-its-kind-strategic-relationship (archived at https://perma.cc/B7HS-ZPUH)
22 M Goldbart. 'Love Island' Lands On Disney+ UK Under Landmark Content Swap Deal With ITV Which Will See 'The Bear' Head To ITVX, *Deadline*, 10 July 2025,

https://deadline.com/2025/07/disney-itv-landmark-deal-love-island-the-bear-the-kardashians-1236453780 (archived at https://perma.cc/KP4S-YU96)

23 J Callaham. Samsung TV Plus: Everything about the free streaming service, Android Authority, 7 October 2024, www.androidauthority.com/samsung-tv-plus-1657262 (archived at https://perma.cc/YPB3-WC2F)

24 T Clark. 'Westworld' is leaving HBO Max entirely after getting canceled, 12 December 2022, www.businessinsider.com/westworld-leaving-hbo-max-after-getting-canceled-reports-2022-12 (archived at https://perma.cc/A64D-RKRR)

25 A B Vary. *Variety*, Warner Bros. Discovery CEO Defends Axing 'Batgirl': We're Not Going to Put a Movie Out Unless We Believe in It, 4 August 2022, https://variety.com/2022/film/news/batgirl-david-zaslav-warner-bros-discovery-1235333681 (archived at https://perma.cc/JML4-YLFW)

26 Discovery to Separate into Two Leading Media Companies, Warner Bros, 21 June 2025 www.wbd.com/news/warner-bros-discovery-separate-two-leading-media-companies (archived at https://perma.cc/2LLE-LLBC)

27 O Darcy and B Stelter. CNN+ will shut down at the end of April, CNN, 21 April 2022 https://edition.cnn.com/2022/04/21/media/cnn-shutting-down (archived at https://perma.cc/XC9U-AEVG)

28 Warner Bros. Discovery to Separate into Two Leading Media Companies, Warner Bros Discovery, 9 June 2025, www.wbd.com/news/warner-bros-discovery-separate-two-leading-media-companies (archived at https://perma.cc/2LLE-LLBC)

29 C Henry. Is there a perfect length for digital video? Digital Content Next, 27 March 2025, https://digitalcontentnext.org/blog/2025/03/27/is-there-a-perfect-length-for-digital-video/ (archived at https://perma.cc/P2PQ-GPHJ)

30 Annual admissions – 1935 onwards, UK Cinema Association, www.cinemauk.org.uk/the-industry/facts-and-figures/uk-cinema-admissions-and-box-office/annual-admissions (archived at https://perma.cc/E6EE-NF9V)

31 UK box office sees further recovery in 2023, UK Cinema Association, 8 January 2024, www.cinemauk.org.uk/2024/01/uk-box-office-sees-further-recovery-in-2023 (archived at https://perma.cc/8HNZ-T9C9)

32 Domestic movie theatrical market summary 1995 to 2025, The Numbers, www.the-numbers.com/market (archived at https://perma.cc/7MEH-7JCK)

33 K Mukherjee. Dwayne Johnson and Chris Evans' Box Office Disaster With 30% RT Becomes Amazon Prime's Biggest Streaming Debut, FandomWire, 17 December 2024 https://fandomwire.com/dwayne-johnson-and-chris-evans-box-office-disaster-with-30-rt-becomes-amazon-primes-biggest-streaming-debut (archived at https://perma.cc/H7BM-GKUH)

34 H Freeman. Tom Hanks on surviving coronavirus: 'I had crippling body aches, fatigue and couldn't concentrate', *The Guardian*, 6 July 2020, www.theguardian.com/film/2020/jul/06/tom-hanks-on-surviving-coronavirus-i-had-crippling-body-aches-fatigue-and-couldnt-concentrate (archived at https://perma.cc/X8DX-DXTJ)

35 C Henry. Tom Hanks is "Actually Thrilled" 'Greyhound' is Coming Out on Apple TV+. No. Really, Mac Observer, 7 December 2020 www.macobserver.com/news/tom-hanks-is-actually-thrilled-greyhound-is-coming-out-on-apple-tv-no-really (archived at https://perma.cc/2B3W-AGRA)

Chapter 3

1. Sky Sports' first Premier League intro, YouTube, 16 August 2022, www.youtube.com/watch?v=WqWZE-iHCaI (archived at https://perma.cc/RP4Q-42DM)
2. S Robson. Sports rights in the US approach $30 billion in 2024, S&P Global, 2 April 2024, www.spglobal.com/market-intelligence/en/news-insights/research/sports-rights-in-the-us-approach-30-billion-in-2024 (archived at https://perma.cc/6FHS-FTA3)
3. J Hussai. Sports Rights: The Jump Ball In The Streaming Ecosystem, S&P Global, 18 June 2024, www.spglobal.com/ratings/en/research/articles/240618-sports-rights-the-jump-ball-in-the-streaming-ecosystem-13151683 (archived at https://perma.cc/QSS4-KA3M)
4. A Rahman. Australia's Optus secures 6-year extension for Premier League Soccer rights, *Hollywood Reporter*, 18 November 2021, www.hollywoodreporter.com/tv/tv-news/optus-premier-league-tv-deal-1235050359 (archived at https://perma.cc/72T7-DML2)
5. M Pegan. Nine buys Premier League rights in Australia as Optus Sport shuts down, *The Guardian*, 30 June 2025 www.theguardian.com/sport/2025/jun/30/nine-entertainment-buys-premier-league-rights-australia-optus-sport-shuts-down (archived at https://perma.cc/3A9P-SUDV)
6. L Reilly. Peacock's exclusive NFL playoff game drove 'biggest subscriber acquisition moment ever measured,' analytics firm says, CNN Business, 24 January 2024, https://edition.cnn.com/2024/01/24/media/peacock-nfl-biggest-subscriber-acquisition-moment-ever-measured (archived at https://perma.cc/CLC9-LUWT)

7 The year of Peacock, 14 November 2024, Comcast, https://corporate.comcast.com/press/releases/the-year-of-peacock-2024#:~:text=2024%20Paris%20Olympic%20Games&text=Fans%20streamed%20an%20incredible%202023.5,streamed%20Games%20in%20U.S.%20history (archived at https://perma.cc/M82A-SVZZ)

8 The gauge: Olympics drive big increases for NBC and Peacock, fuel above average TV viewership in August, Nielson, September 2024, www.nielsen.com/news-center/2024/the-gauge-olympics-drive-big-increases-for-nbc-and-peacock-fuel-above-average-tv-viewership-in-august (archived at https://perma.cc/S6YR-QRKC)

9 Ibid.

10 M Sweney. Premier League brings record number of sign-ups to Amazon Prime, *The Guardian*, 6 December 2019, www.theguardian.com/technology/2019/dec/06/premier-league-brings-record-number-of-sign-ups-to-amazon-prime (archived at https://perma.cc/6QAT-EV2Y)

11 J Kanter. Amazon suffers scattered issues on first night of Premier League coverage in the UK, Deadline, 4 December 2019, https://deadline.com/2019/12/amazon-scattered-issues-premier-league-1202800177/ (archived at https://perma.cc/P82D-2653)

12 Netflix and most valuable promotions' Jake Paul vs Mike Tyson mega-event makes history with over 108 million live global viewers, Netflix, 19 November 2024, https://about.netflix.com/en/news/jake-paul-vs-mike-tyson-over-108-million-live-global-viewers (archived at https://perma.cc/PXP6-4BJN)

13 T Spangler. Beyoncé NFL Halftime Show Will Be Available to Rewatch on Netflix as a Stand-Alone Special, *Variety*, 25 December 2024 https://variety.com/2024/tv/news/beyonce-nfl-halftime-show-christmas-netflix-rewatch-1236258011 (archived at https://perma.cc/JVG7-KWJB)

14 DAZN founder Len Blavatnik named in Sports Illustrated's 'Most Influential People in Boxing' list, DAZN, 17 December 2024, www.dazn.com/en-GB/news/boxing/dazn-founder-len-blavatnik-named-to-sports-illustrateds-most-influential-people-in-boxing-list/1l5f0x3vnxnfh1u05b5urxer2h (archived at https://perma.cc/FK4D-ENMV)

15 D Thomas. Len Blavatnik pumps further $800mn into creating 'Spotify of sport', *Financial Times*, 23 January 2025, www.ft.com/content/4aff9714-f70f-4ce1-ac8d-f2fcf682a434 (archived at https://perma.cc/2R4K-TUC6)

16 DAZN acquires exclusive live broadcast rights to UEFA competitions in New Zealand, DAZN, 6 August 2024, https://dazngroup.com/press-room/dazn-acquires-exclusive-live-broadcast-rights-to-uefa-competitions-in-new-zealand (archived at https://perma.cc/YT44-TVQ3)

17 R Bresaola. DAZN Canada retains rights to UEFA Champions League, Sportcal, 16 January 2024, www.sportcal.com/media/dazn-canada-retains-rights-to-uefa-champions-league (archived at https://perma.cc/2H7C-GZ9B)

18 M Smiley. The 'Netflix of Sport' just acquired AFL and NRL rights. Meet the curious billionaire behind it all, ABC News, 7 May 2025, www.abc.net.au/news/2025-05-08/len-blavatnik-dazn-warner-foxtel-nrl-afl-soviet-oligarch/105253268 (archived at https://perma.cc/S6LP-48JT)

19 DAZN and Ligue 1 break up: League considers going DTC, Enders Analysis, 12 May 2025, www.endersanalysis.com/reports/dazn-and-ligue-1-break-league-considers-going-dtc (archived at https://perma.cc/SKF3-4LM6)

20 Ligue 1 goes direct-to-consumer: Partners needed, Enders Analysis, 9 July 2025 www.endersanalysis.com/reports/

ligue-1-goes-direct-consumer-partners-needed#:~:text=After%20four%20failed%20broadcast%20licence,arrangements%20with%20third%2Dparty%20platforms (archived at https://perma.cc/ST8G-WYRD)

21 D Moore. Streamers will spend $12.5bn on sports rights in 2025, led by DAZN, Ampere Analysis, 18 February 2025, https://ampereanalysis.com/insight/streamers-will-spend-125bn-on-sports-rights-in-2025-led-by-dazn (archived at https://perma.cc/E8FG-5BPF)

22 M Sweney. Football on TV: fans 'paying almost 60% more to watch all big games than in 2020', *The Guardian*, 12 January 2025, www.theguardian.com/football/2025/jan/12/football-on-tv-fans-paying-almost-60-more-to-watch-all-big-games-than-in-2020 (archived at https://perma.cc/5252-7ZUT)

23 T Garry. Women's Super League viewing figures soar after streaming switch to YouTube, *The Guardian*, 10 October 2024, www.theguardian.com/football/2024/oct/10/womens-super-league-football-youtube-streaming (archived at https://perma.cc/NR83-MSN2)

24 H Goldblatt. Goals for 2027, 2031: More FIFA Women's Soccer on Netflix, Netflix, 8 August 2025 www.netflix.com/tudum/articles/womens-world-cup-netflix (archived at https://perma.cc/QVX9-FJES)

25 T Garry. WSL agrees record £65m domestic five-year TV deal with Sky Sports and BBC, *The Guardian*, 30 October 2024 www.theguardian.com/football/2024/oct/30/wsl-agrees-record-65m-domestic-five-year-tv-deal-with-sky-sports-and-bbc (archived at https://perma.cc/QRU6-MC3U)

26 S Custis. Prince William reveals he uses the KIDS as part of Villa matchday rituals & how he lurks on fan forums with random name, *The Sun*, 13 March 2025, www.thesun.co.uk/

sport/33843926/prince-william-aston-villa-match-superstitions (archived at https://perma.cc/S2KY-RAKA)

27 S Wrack. WSL frustrated in attempts to broadcast games during 3pm Saturday black-out, *The Guardian*, 11 September 2024, www.theguardian.com/football/article/2024/sep/11/wsl-broadcast-saturday-blackout-womens-football (archived at https://perma.cc/TCX6-ZHET)

28 A Marchand. New ESPN, Fox, WBD streaming venture won't solve much – at least not yet, *New York Times*, 7 February 2024, www.nytimes.com/athletic/5256834/2024/02/07/espn-fox-warner-sports-streaming-service-marchand (archived at https://perma.cc/N2ZM-4ZVA)

29 O Coryell. Pricing confirmed at $42.99/month for upcoming Venu sports streaming service, ESPN Press Room, 1 August 2024, https://espnpressroom.com/us/press-releases/2024/08/pricing-confirmed-at-42-99-month-for-upcoming-venu-sports-streaming-service (archived at https://perma.cc/X9Q8-MUV9)

30 A Marchand. New ESPN, Fox, WBD streaming venture won't solve much – at least not yet, *New York Times*, 7 February 2024, www.nytimes.com/athletic/5256834/2024/02/07/espn-fox-warner-sports-streaming-service-marchand (archived at https://perma.cc/N2ZM-4ZVA)

31 T DeMeyer and A Marchand, Venu Sports launch temporarily blocked following Fubo's legal challenge, *New York Times*, 16 August 2024, www.nytimes.com/athletic/5705694/2024/08/16/sports-streaming-disney-warner-fox-venu-judge (archived at https://perma.cc/8RXF-LCCU)

32 ESPN's direct-to-consumer streaming service set for fall launch, ESPN, 13 May 2025 www.espn.co.uk/espn/story/_/id/45126967/espn-direct-consumer-streaming-service-set-fall-launch (archived at https://perma.cc/5YX3-GGNS)

Chapter 4

1. About Twitch, www.twitch.tv/p/en/about (archived at https://perma.cc/Z8WT-LB87)
2. J Siegel and C Hamilton. YouTube is going LIVE, YouTube Official Blog, 8 April 2011, https://blog.youtube/news-and-events/youtube-is-going-live (archived at https://perma.cc/5DDJ-NC86)
3. Forbes, Tyler Blevins (Ninja), 26 September 2023, www.forbes.com/profile/tyler-blevins-ninja (archived at https://perma.cc/3328-NRBL)
4. Twitch, Amazon ads, https://advertising.amazon.com/en-gb/channels/twitch (archived at https://perma.cc/A3T5-G2VB)
5. Ibid.
6. What we watched: A Netflix Engagement Report, Netflix, 12 December 2023, https://about.netflix.com/en/news/what-we-watched-a-netflix-engagement-report (archived at https://perma.cc/48K2-ERFH)
7. Global Live Streaming Market 2024–2033, Custom Market Insights, www.custommarketinsights.com/report/live-streaming-market (archived at https://perma.cc/76RX-6RUM)
8. Video viewing forecasts: A slowdown in change, Enders Analysis, 5 August 2024, www.endersanalysis.com/reports/video-viewing-forecasts-slowdown-change (archived at https://perma.cc/YR8V-BZ7S)
9. K Gittleson. 13 Amazon buys video-game streaming site Twitch, BBC, 25 August 2014 www.bbc.co.uk/news/technology-28930781 (archived at https://perma.cc/Y7PV-D5MK)
10. S Rodriguez. Amazon paid almost $1 billion for Twitch in 2014. It's still losing money, *The Wall Street Journal*, 29 July 2024, www.wsj.com/tech/twitch-amazon-video-games-investment-9020db87 (archived at https://perma.cc/FU93-JNFG)

11 J Siegel and C Hamilton. YouTube is going LIVE, YouTube Official Blog, 8 April 2011, https://blog.youtube/news-and-events/youtube-is-going-live (archived at https://perma.cc/5DDJ-NC86)

12 G Park. YouTube signs on exclusive streaming deals with 3 big gaming creators: LazarBeam, Muselk and Valkyrae, 13 January 2020 www.washingtonpost.com/video-games/2020/01/13/youtube-signs-exclusive-streaming-deals-with-3-big-gaming-creators-lazarbeam-muselk-valkyrae (archived at https://perma.cc/TA7U-VK5Y)

13 K Gordon. Burnout turned Twitch streamers' dreams of playing games full time into nightmares, NPR, 16 August 2022, www.npr.org/2022/08/16/1117650184/twitch-streamers-burnout-video-games (archived at https://perma.cc/ME58-Q86Y)

14 K MacDonald. 'I am not gonna die on the internet for you!': how game streaming went from dream job to a burnout nightmare, *The Guardian*, 26 November 2021, www.theguardian.com/games/2021/nov/26/i-am-not-gonna-die-on-the-internet-for-you-how-game-streaming-went-from-dream-job-to-a-burnout-nightmare (archived at https://perma.cc/D77S-QQVX)

15 D Harwell. Inside the life of a 24/7 streamer: 'What more do you want?', *The Washington Post*, 4 May 2025, www.washingtonpost.com/technology/2025/05/04/longest-marathon-streamer-emilycc (archived at https://perma.cc/2SSH-P64A)

16 Zoom Video Communications, Q1 FY21 Earnings, Zoom, 2 June 2020, https://investors.zoom.us/static-files/32469119-f0a2-4e77-a2f8-45d97032e40e (archived at https://perma.cc/EE4B-VEVZ)

17 B Logan. Comedian Catherine Bohart: 'When Covid hit it was, "there goes my next six months earnings"', *The*

Guardian, 3 February 2022, www.theguardian.com/culture/2022/feb/03/comedian-catherine-bohart-when-covid-hit-it-was-there-goes-my-next-six-months-earnings (archived at https://perma.cc/3GCG-KN2B)

18 Ibid.

19 E Farwell. 'It Was Like Performing for the First Time': 11 comedians reflect on what they've learned from months of Zoom and outdoor shows during COVID-19, Vulture, 23 December 2020, www.vulture.com/article/comedians-zoom-outdoor-shows-covid.html (archived at https://perma.cc/9Z6M-GLAU)

20 A Frank. How 'quarantine concerts' are keeping live music alive as venues remain closed, Vox, 8 April 2020, www.vox.com/culture/2020/4/8/21188670/coronavirus-quarantine-virtual-concerts-livestream-instagram (archived at https://perma.cc/96DV-WTN7)

21 Ibid.

22 L McMahon. What it's really like to be a female Twitch streamer – and why it's far from game over for misogyny in the gaming industry, *The Scotsman*, 8 August 2021, www.scotsman.com/news/people/what-its-really-like-to-be-a-female-twitch-streamer-3337578 (archived at https://perma.cc/U66J-SLJ7)

23 S Powell. Twitch: Streamers call for a blackout to recognise victims of sexual and racial abuses, BBC News, 24 June 2020, www.bbc.co.uk/news/newsbeat-53164395 (archived at https://perma.cc/DS3T-EENF)

24 K Anciones and M Checa-Romero. Sexualized culture on livestreaming platforms: a content analysis of Twitch.tv (archived at https://perma.cc/YQ22-XHCF), *Humanities and Social Sciences Communications* (2024) 11(1)

25 An update to our attire policy, Twitch, 3 January 2024, https://safety.twitch.tv/s/article/An-Update-to-Our-Attire-

Policy?language=en_US (archived at https://perma.cc/V2V3-799T)

26 S Leitch, The making of a 'made for social media massacre', in *Rethinking Social Media and Extremism* (2022), S Leitch and P Pickering, ANU Press; T Haselton and M Graham, About 2,200 people watched the German synagogue shooting on Amazon's Twitch, CNBC, 9 October 2019, www.cnbc.com/2019/10/09/the-german-synagogue-shooting-was-streamed-on-twitch.html (archived at https://perma.cc/DS9H-A5C2); M Graham, Livestreamed shootings have advertisers demanding better safety from sites like Facebook and YouTube, CNBC, 9 October 2019, www.cnbc.com/2019/10/16/advertisers-on-twitch-facebook-grapple-with-safety-of-live-streaming.html (archived at https://perma.cc/Y28F-3PPT)

27 Big Tech Backslide: How 7 Social-Media Rollbacks Endanger Democracy Ahead of the 2024 Elections, December 2023 www.freepress.net/big-tech-backslide-how-social-media-rollbacks-endanger-democracy-ahead-2024-elections (archived at https://perma.cc/FP2Z-9RMK)

28 @ryanstoker, TikTok, 20 July 2022, www.tiktok.com/@ryanstoker/video/7122480674967162117?lang=en (archived at https://perma.cc/QPQ8-AWNV)

Chapter 5

1 E Forde. Oversharing: how Napster nearly killed the music industry, *The Guardian*, 31 May 2019 www.theguardian.com/music/2019/may/31/napster-twenty-years-music-revolution (archived at https://perma.cc/RTC4-DGPC)

2 J Clover. Apple Officially Splits iTunes for Windows Into Apple Music, TV, and Devices Apps, Mac Rumours, 7 February 2024 www.macrumors.com/2024/02/07/apple-splits-itunes-for-windows (archived at https://perma.cc/3HRT-QD3T)

3 W Isaacson. Steve Jobs: The exclusive biography, Abacus, 2011

4 H McIntyre. What exactly is stream-ripping, the new way people are stealing music, Forbes, 11 August 2017, www.forbes.com/sites/hughmcintyre/2017/08/11/what-exactly-is-stream-ripping-the-new-way-people-are-stealing-music (archived at https://perma.cc/BZQ6-BYJX)

5 www.macrumors.com/2024/02/07/apple-splits-itunes-for-windows (archived at https://perma.cc/3HRT-QD3T)

6 United States Securities and Exchange Commission, Form 10-K/A (Amendment No. 1), Annual Report Pursuant to Section 13 or 15(d) of The Securities Exchange Act of 1934, 26 September 2009, www.sec.gov/ix?doc=/Archives/edgar/data/320193/000032019319000093/a8-kseptember201991019.htmwww.sec.gov/Archives/edgar/data/320193/000119312510012091/d10ka.htm (archived at https://perma.cc/7ZYB-QDLC)

7 Introducing Apple Music – All The Ways You Love Music. All in One Place, Apple, 8 June 2015 www.apple.com/uk/newsroom/2015/06/08Introducing-Apple-Music-All-The-Ways-You-Love-Music-All-in-One-Place

8 R Neate. Daniel Ek profile: 'Spotify will be worth tens of billions', *The Telegraph*, 17 February 2010, www.telegraph.co.uk/finance/newsbysector/mediatechnologyandtelecoms/media/7259509/Daniel-Ek-profile-Spotify-will-be-worth-tens-of-billions.html (archived at https://perma.cc/95AL-S8DT)

9 Spotify Launches in India, Spotify, 26 February 2019 https://newsroom.spotify.com/2019-02-26/spotify-launches-in-india (archived at https://perma.cc/FLC7-6AE7)

10 J Weatherbed. Spotify finally turned a profit for a full year, The Verge, 4 February 2025, www.theverge.com/

news/605709/spotify-earnings-q4-2024-profitability (archived at https://perma.cc/JYW4-LK8Z)

11 A Steele. Joe Rogan gets new Spotify deal worth up to $250 million, The Wall Street Journal, 2 February 2024, www.wsj.com/business/media/joe-rogan-podcast-spotify-deal-28eb5f74 (archived at https://perma.cc/K464-8MUM)

12 Everything In Store for Podcast Listeners and Creators This International Podcast Day, Spotify, 28 September 2023 https://newsroom.spotify.com/2023-09-28/international-podcast-day-transcripts-chapters-show-pages-global/ (archived at https://perma.cc/23DZ-2J8S)

13 About Spotify, https://newsroom.spotify.com/company-info/ (archived at https://perma.cc/DY5Y-GDUG)

14 Spotify Expands Its Audiobooks Offering To Listeners In Germany, Austria, Switzerland, and Liechtenstein, Spotify, 15 April 2025 https://newsroom.spotify.com/2025-04-15/spotify-expands-its-audiobooks-offering-to-listeners-in-germany-austria-switzerland-and-liechtenstein/ (archived at https://https://perma.cc/5BWV-M9W9)

15 Music videos rolling out in beta to Premium Spotify users across select markets, For the Record, Spotify, 13 March 2024, https://newsroom.spotify.com/2024-03-13/music-videos-rolling-out-in-beta-to-premium-spotify-users-across-select-markets (archived at https://perma.cc/M9BT-6LGW)

16 250,000 Video Podcasts and Counting: Creators and Audiences Are Embracing Video Content on Spotify, Spotify, 28 June 2024 https://newsroom.spotify.com/2024-06-28/250000-video-podcasts-and-counting-creators-and-audiences-are-embracing-video-content-on-spotify/ (archived at https://perma.cc/K38P-VRBT)

17 Spotify Tests Video-Based Learning Courses in the UK, Spotify, 25 March 2024 https://newsroom.spotify.com/

2024-03-25/spotify-tests-video-based-learning-courses-in-the-uk/ (archived at https://perma.cc/8SMW-Q6ZL)

18 How He Built This: Hear Don Katz on the Amazing Story of Audible's Founding, Audible, 2 November 2021 https://www.audible.com/about/newsroom/how-he-built-this-hear-don-katz-on-the-amazing-story-of-audibles-founding (archived at https://perma.cc/XWX3-FEUF)

19 K Fairbrother. Surprise! The UK audiobook market stands at over £1bn, The Bookseller, 19 October 2024, www.thebookseller.com/features/surprise-the-uk-audiobook-market-stands-at-over-1bn (archived at https://perma.cc/VX7H-B6JP)

20 Our strategy and history, Deezer https://www.deezer-investors.com (archived at https://perma.cc/MK8T-VU8J)

21 M Dalugdug. YouTube Music emerges as 'most adopted music streaming service' in Q2, Kantar, 20 August 2024, www.musicbusinessworldwide.com/youtube-music-emerges-as-most-adopted-music-streaming-service-in-q2-kantar (archived at https://perma.cc/XYC9-HU5W)

22 It's time to fix streaming, Musicians' Union, https://musiciansunion.org.uk/campaigns/fix-streaming-and-keep-music-alive (archived at https://perma.cc/F3FJ-NLBR)

23 L Fisher. Paul McCartney and Chris Martin lead 150 artists demanding reform to music streaming laws, *The Telegraph*, 19 April 2021, www.telegraph.co.uk/news/2021/04/19/paul-mccartney-chris-martin-lead-150-artists-demanding-reform (archived at https://perma.cc/7G7F-SVWY)

24 Spotify, Our Annual Music Economics Report: A record $10 billion payout, March 2025, https://loudandclear.byspotify.com/#takeaway-1 (archived at https://perma.cc/R4XU-5WTP)

25 Royalties, Spotify, https://support.spotify.com/us/artists/article/royalties (archived at https://perma.cc/BB97-XCJU)

26 Our Annual Music Economics Report: Streaming's rising tide, Spotify, March 2025, https://loudandclear.byspotify.com/#takeaway-2 (archived at https://perma.cc/7ZQL-HQ48)

27 2024 Music Economics Report, Duetti, https://report.duetti.co/#6 (archived at https://perma.cc/UT4B-XP3X)

28 A Paine. Ahead of Record Store Day, Official Charts Company marks first decade of vinyl albums chart, 7 April 2025 www.musicweek.com/labels/read/ahead-of-record-store-day-official-charts-company-marks-first-decade-of-vinyl-albums-chart/091717 (archived at https://perma.cc/9S9Y-2K4Y)

29 B Hammersley. Audible revolution, *The Guardian*, 12 February 2004, www.theguardian.com/media/2004/feb/12/broadcasting.digitalmedia (archived at https://perma.cc/W9EV-7XQ8)

30 A Shapiro. The iPod is dead, but the podcast lives on, The Verge, 15 May 2022, www.theverge.com/2022/5/15/23071515/ipod-dead-podcast-legacy-apple-spotify (archived at https://perma.cc/Y4QG-5C5M)

31 B Hammersley. Audible revolution, *The Guardian*, 12 February 2004, www.theguardian.com/media/2004/feb/12/broadcasting.digitalmedia (archived at https://perma.cc/W9EV-7XQ8)

32 S Singh. Podcast statistics 2025 – number of podcasts, listeners & trends, Demandsage, 6 May 2025, www.demandsage.com/podcast-statistics (archived at https://perma.cc/24PK-X9MF)

33 A Shapiro. The iPod is dead, but the podcast lives on, The Verge, 15 May 2022, www.theverge.com/2022/5/15/23071515/ipod-dead-podcast-legacy-apple-spotify (archived at https://perma.cc/Y4QG-5C5M)

34 M Rothblatt. Meet the new biotech billionaire putting pig hearts in humans, Forbes, 18 July 2024, www.forbes.com/sites/richardjchang/2024/07/18/meet-the-new-biotech-martine-rothblatt-billionaire-putting-pig-hearts-in-humans (archived at https://perma.cc/3U4V-K7DA)

35 2025 Annual Meeting of Stockholders, SiriusXM, 28 May 2025, https://investor.siriusxm.com (archived at https://perma.cc/29J2-WPFB)

36 SiriusXM reports first quarter 2025 operating and financial results, SiriusXM, 1 May 2025, https://d1io3yog0oux5.cloudfront.net/_55ce2e0d390679303bb94dfccd9b6bd9/siriusxm/db/2244/21504/earnings_release/SIRI+Q1+2025+Earnings+Release+-+PUBLICvF.pdf (archived at https://perma.cc/T4ZH-Y2Z9)

37 J Schott. Apple makes groundbreaking deal with TuneIn to expand radio offerings, Fox Business, 25 September 2019, www.foxbusiness.com/technology/apple-makes-groundbreaking-deal-with-tunein-to-expand-radio-offerings (archived at https://perma.cc/FG33-B7S9)

Chapter 6

1 D&P Advisory, The Crystal Ball – IPL Prediction 2024, https://dandpadvisory.com/the-crystal-ball-ipl-prediction-2024 (archived at https://perma.cc/KJJ3-EGPX)

2 IPL media rights sold in record-breaking $6bn deal, BBC News, 15 June 2022, www.bbc.co.uk/news/world-asia-india-61793888 (archived at https://perma.cc/256T-47TL)

3 S Bhattacharjee. India vs Pakistan, Champions Trophy 2025: India-Pak clash breaks viewership records with 61+ crore views on JioHotstar, 24 February 2025, www.mykhel.com/cricket/india-vs-pakistan-champions-trophy-2025-india-pak-clash-breaks-viewership-records-with-61-crore-views-on-jiohotstar-343188.html (archived at https://perma.cc/4ERP-MPS6)

4 Number of smartphone users in India in 2010 to 2023, with estimates until 204018 September 2023, Statista, www.statista.com/statistics/467163/forecast-of-smartphone-users-in-india (archived at https://perma.cc/W4NQ-R38Z)

5 TOI Tech Desk, JioHotstar subscription plans explained: pricing, features and other details, *Times of India*, 16 February 2025, https://timesofindia.indiatimes.com/technology/tech-news/jiohotstar-subscription-plans-explained-pricing-features-and-all-other-details/articleshow/118241839.cms (archived at https://perma.cc/Z9FX-BK77)

6 W Ma. Apple held talks With China Mobile to bring Apple TV+ to China, The Information, 4 June 2024, www.theinformation.com/articles/apple-held-talks-with-china-mobile-to-bring-apple-tv-to-china (archived at https://perma.cc/RJR2-EV4H)

7 W Ma. Inside Tim Cook's secret $275 billion deal with Chinese authorities, The Information, 7 December 2021, www.theinformation.com/articles/facing-hostile-chinese-authorities-apple-ceo-signed-275-billion-deal-with-them (archived at https://perma.cc/HLA2-6NE4)

8 M Vengattil. Apple moving to make most iPhones for US in India rather than China, source says, Reuters, 25 April 2025, www.reuters.com/world/china/apple-aims-source-all-us-iphones-india-pivot-away-china-ft-reports-2025-04-25 (archived at https://perma.cc/K5RL-NXMR)

9 O Zhao. As Chinese SVoDs expand overseas, can Chinese content find an audience? Ampere Analysis, 11 October 2023, https://ampereanalysis.com/insight/as-chinese-svods-expand-overseas-can-chinese-content-find-an-audience (archived at https://perma.cc/9NB7-NLTZ)

10 N Mishra. Tencent and Iqiyi lead China's streaming revolution, campaign, 3 October 2024, www.campaignasia.com/article/tencent-and-iqiyi-lead-chinasstreaming-revolution/498671 (archived at https://perma.cc/B9GS-9ZHG)

11 Online views of the Paris 2024 Olympics on leading streaming apps in China, by sport, Statista, 20 May 2025, www.statista.com/statistics/1552112/china-online-viewership-of-paris-2024-olympics-by-sport (archived at https://perma.cc/V8Z3-H57N)

12 K Moskvitch. China bans Tiananmen Square-related web search terms, BBC News, 4 June 2012, www.bbc.com/news/technology-18321548 (archived at https://perma.cc/ZL9R-QDVL)

13 China bans BBC World News from broadcasting, BBC News, 12 February 2021, www.bbc.com/news/world-asia-china-56030340 (archived at https://perma.cc/SN6D-3ESW)

14 Ofcom revokes CGTN's licence to broadcast in the UK, Ofcom, 4 February 2021, www.ofcom.org.uk/tv-radio-and-on-demand/broadcast-standards/ofcom-revokes-cgtn-licence-to-broadcast-in-uk (archived at https://perma.cc/TA98-X32T)

15 B Dayo. Netflix is launching in Nigeria, but not everyone is happy, VICE, 28 December 2020, www.vice.com/en/article/netflix-naija-launch-nigeria (archived at https://perma.cc/QK9L-LEXQ)

16 Netflix presents a preview of its South African productions and partnerships at MIP Africa, 4 September 2023, Netflix News, https://about.netflix.com/en/news/netflix-presents-a-preview-of-its-south-african-productions-and-partnerships (archived at https://perma.cc/6DMR-BJB8)

17 N Fraser. New government supports for the ABC, Parliament of Australia, 13 February 2025, www.aph.gov.au/About_Parliament/Parliamentary_departments/Parliamentary_Library/Research/FlagPost/2025/February/

New_government_supports_for_the_ABC (archived at https://perma.cc/U7YJ-5U9W)
18 Top Streaming Services by Subscribers, FlixPatrol, https://flixpatrol.com/streaming-services/subscribers (archived at https://perma.cc/2NVV-SUK4)
19 O Zhao. As Chinese SVoDs expand overseas, can Chinese content find an audience? Ampere Analysis, 11 October 2023, https://ampereanalysis.com/insight/as-chinese-svods-expand-overseas-can-chinese-content-find-an-audience (archived at https://perma.cc/9NB7-NLTZ)

Chapter 7

1 B Agate. Streaming units are beginning to be profitable. Here's why, Forbes, 11 October 2024, www.forbes.com/sites/bradadgate/2024/10/11/streaming-units-are-starting-beginning-to-be-profitable (archived at https://perma.cc/8FT9-4FNS)
2 C Widener et al. 2025 Digital Media Trends: Social platforms are becoming a dominant force in media and entertainment, Deloitte, 25 March 2025, www2.deloitte.com/us/en/insights/industry/technology/digital-media-trends-consumption-habits-survey.html (archived at https://perma.cc/DH9U-9C6M)
3 A Canal. Streaming turns a corner as Disney, Paramount report profits – but that doesn't solve all of media's problems, Yahoo! Finance, 17 August 2024, https://finance.yahoo.com/news/streaming-turns-a-corner-as-disney-paramount-report-profits--but-that-doesnt-solve-all-of-medias-problems-153116985.html (archived at https://perma.cc/YM7Q-NEC9)
4 Compared to ten years ago, do you think subscription based video streaming services are better or worse, or are they about the same? YouGov, 11 March 2025, https://yougov.co.uk/topics/technology/survey-results/daily/2025/03/11/4c191/3 (archived at https://perma.cc/HC8E-2DZ7)

5 2024 TV advertising: fact vs. fiction – Wave 7, HUB Entertainment Research, July 2024, https://hubresearchllc.com/reports/?category=2024&title=2024-tv-advertising-fact-vs-fiction-wave-7 (archived at https://perma.cc/FB5V-JD4R)

6 M Sweney. Market for TV streaming advertising to pass £1bn, *The Guardian*, 29 December 2024, www.theguardian.com/media/2024/dec/29/market-for-tv-streaming-advertising-to-pass-1bn-milestone (archived at https://perma.cc/7E3S-E5AX)

7 D Segal. A kid's show juggernaut that leaves nothing to chance, *The New York Times*, 5 May 2022, www.nytimes.com/2022/05/05/arts/television/cocomelon-moonbug-entertainment.html (archived at https://perma.cc/8A7M-7YGM)

8 J Tolentino. How CoComelon Captures Our Children's Attention, 17 June 2024 www.newyorker.com/magazine/2024/06/17/cocomelon-children-television-youtube-netflix (archived at https://perma.cc/YN6D-7GRQ)

9 J Tolentino. How CoComelon captures our children's attention, *The New Yorker*, 10 June 2024, www.newyorker.com/magazine/2024/06/17/cocomelon-children-television-youtube-netflix (archived at https://perma.cc/YN6D-7GRQ)

Chapter 8

1 C Stryker and E Kavlakoglu. What is artificial intelligence (AI)? IBM, 9 August 2024, www.ibm.com/think/topics/artificial-intelligence (archived at https://perma.cc/9CRJ-F6ZJ)

2 Amazon 2024 Shareholder Letter, https://s2.q4cdn.com/299287126/files/doc_financials/2025/ar/2024-Shareholder-Letter-Final.pdf (archived at https://perma.cc/GTK8-WEBQ)

3 A Pulver. The Brutalist and Emilia Perez's voice-cloning controversies make AI the new awards season battleground, *The Guardian*, 20 Jan 2025 www.theguardian.com/film/2025/jan/20/the-brutalist-and-emilia-perezs-voice-cloning-

controversies-make-ai-the-new-awards-season-battleground (archived at https://perma.cc/8QXS-YZ3X)
4. Sora is here, 9 December 2024, OpenAI, https://openai.com/index/sora-is-here/ (archived at https://perma.cc/Q3JR-4U2G)
5. W Cho. OpenAI's Sora, the text-to-video tool that caught Hollywood off guard, rolls out to public, *Hollywood Reporter*, 9 December 2024, www.hollywoodreporter.com/business/business-news/openai-sora-hollywood-public-1236081753 (archived at https://perma.cc/X26X-Z2PG)
6. J Flannery O'Connor and E Moxley. Our approach to responsible AI innovation, YouTube Official Blog, 14 November 2023, https://blog.youtube/inside-youtube/our-approach-to-responsible-ai-innovation (archived at https://perma.cc/Q6LA-C6XM)
7. C Bharadwaj. How AI in media and entertainment industry is revolutionizing the sector, Appinventiv, 25 September 2024, https://appinventiv.com/blog/ai-in-media-and-entertainment (archived at https://perma.cc/3ESZ-JHXY)
8. C Johnson. With Play Something, Netflix does all the work for you, Netflix News, 28 April 2021, https://about.netflix.com/en/news/play-something-netflix-does-the-work-for-you (archived at https://perma.cc/QUT8-ZR8E)
9. A R Gonçalvesa, D Costa Pinto, S Shuqair, M Dalmoro and A S Mattila. Artificial intelligence vs. autonomous decision-making in streaming platforms: A mixed-method approach, *International Journal of Information Management*, 76, 102748, June 2024
10. Unveiling our innovative new TV experience featuring enhanced design, responsive recommendations and a new way to search, Netflix News, 7 May 2025, https://about.netflix.com/en/news/unveiling-our-innovative-new-tv-experience (archived at https://perma.cc/CJH5-GPAD)

11 Deezer deploys cutting-edge AI detection tool for music streaming, Deezer, 24 January 2025, https://newsroom-deezer.com/2025/01/deezer-deploys-cutting-edge-ai-detection-tool-for-music-streaming (archived at https://perma.cc/2W9F-FDKE)

12 Spotify debuts a new AI DJ, right in your pocket, For the Record, Spotify, 22 February 2023, https://newsroom.spotify.com/2023-02-22/spotify-debuts-a-new-ai-dj-right-in-your-pocket (archived at https://perma.cc/SUW9-CZ3N)

13 Spotify to acquire Sonantic, an AI voice platform, For the Record, Spotify, 13 June 2022, https://newsroom.spotify.com/2022-06-13/spotify-to-acquire-sonantic-an-ai-voice-platform (archived at https://perma.cc/6NQP-2QMW)

14 Fake Obama created using AI tool to make phoney speeches, BBC News, 17 July 2017, www.bbc.co.uk/news/av/technology-40598465 (archived at https://perma.cc/7NC9-NA8G)

15 M M Grynbaum and R Mac. The Times sues OpenAI and Microsoft over A.I. use of copyrighted work, *New York Times*, 27 December 2023, www.nytimes.com/2023/12/27/business/media/new-york-times-open-ai-microsoft-lawsuit.html (archived at https://perma.cc/C9X7-M3MH)

16 The Times and Amazon Announce an A.I. Licensing Deal, New York Times, 29 May 2025 https://www.nytimes.com/2025/05/29/business/media/new-york-times-amazon-ai-licensing.html (archived at https://perma.cc/9YH9-KPPH)

17 Runway Partners with Lionsgate in First-of-its-Kind AI Collaboration, Business Wire, 18 September 2024, www.businesswire.com/news/home/20240918322724/en/Runway-Partners-with-Lionsgate-in-First-of-its-Kind-AI-Collaboration (archived at https://perma.cc/EJ4V-VY8J)

18 T Conlan. Streaming demand for UK shows will create 30,000 film and TV jobs, *The Guardian*, 12 September

2021, www.theguardian.com/media/2021/sep/12/ streaming-demand-for-uk-shows-will-create-30000-film-and-tv-jobs (archived at https://perma.cc/Q95G-UU8H)
19 Ibid.
20 W Cho. The Hollywood jobs most at risk from AI, *Hollywood Reporter*, 30 January 2024, www.hollywoodreporter.com/business/business-news/ai-hollywood-workers-job-cuts-1235811009 (archived at https://perma.cc/4R4B-4JDT)
21 S Mahanty. Emma Corrin: 'I feel very settled in myself, which I haven't had for a long time', *Elle*, 10 April 2025, www.elle.com/uk/life-and-culture/culture/a64320637/emma-corrin-interview (archived at https://perma.cc/Z8TM-DVG5)
22 J McHenry. The best shows on Netflix to use as background noise, Vulture, 6 March 2018, www.vulture.com/2018/03/best-netflix-shows-background-noise.html (archived at https://perma.cc/ER4G-A22K)
23 J A Starosta and B Izydorczyk. Understanding the phenomenon of binge-watching – a systematic review, *International Journal of Environmental Research and Public Health*, 17(12), 4469, 22 June 2020

Conclusion

1 D Anguiano. Sag-Aftra union ratifies strike-ending contract with Hollywood studios, *The Guardian*, 6 December 2023, www.theguardian.com/culture/2023/dec/05/sag-aftra-union-ratifies-contract-hollywood-studios (archived at https://perma.cc/UHM2-Z6UA)
2 D Anguiano. Film and TV crew members reach deal with Hollywood studios to avert strike, *The Guardian*, 26 June 2024, www.theguardian.com/film/article/2024/jun/26/iatse-union-hollywood-deal (archived at https://perma.cc/5WUP-J9KA)

3 M R Hasan, A K Jha and Y Liu. Excessive use of online video streaming services: Impact of recommender system use, psychological factors, and motives, *Computers in Human Behavior*, 14 November 2017, https://mural.maynoothuniversity.ie/id/eprint/15956/1/RH_excessive.pdf (archived at https://perma.cc/D6MF-LPRG)

4 T Clark. How major Hollywood studios are shifting their streaming strategies as the theater industry stages a comeback, Business Insider, 11 May 2022, www.businessinsider.com/how-long-movies-play-in-theaters-before-streaming-2022-5 (archived at https://perma.cc/H78R-8SDG)

5 Cinema United, Cinemacon state of the industry: Cinema United president and CEO Michael O'Leary calls upon exhibition and distribution to come together to transform industry, 1 April 2025, https://cinemaunited.org/2025/04/01/cinemacon-state-of-the-industry-cinema-united-president-and-ceo-michael-oleary-calls-upon-exhibition-and-distribution-to-come-together-to-transform-industry/ (archived at https://perma.cc/8RAM-RALL)

INDEX

5G internet 110
8Chan 123, 124
9Now 161

ABC iview 160
Adolescence 187
advertising 170–72
  on Disney+ 10
  during live sport 32, 81–83, 85–86, 170
  free ad-supported television (FAST) 67–68
  on Netflix 10, 22–25, 28, 32, 86
  on Peacock 85
  on podcasts 23
  on Prime Video 23, 86, 170
  on Spotify 134
  on Twitch 113
  on YouTube 23, 112–13
Alibaba 155
Amazon 2, 6, 43–50, 60–61, 63, 163
  AI use 176–77
  Alexa 139
  Amazon Music 138
    royalties paid 143
  Amazon Web Services (AWS) 50, 63
  Audible 136, 137–38
  Fire Stick 68, 77
  MGM, purchase of 46–48
  password sharing 169
  Prime Video 9, 13, 26, 45–50, 52, 75, 195
    advertising on 23, 86, 170
    on Apple TV 62
    Channels 61, 72
    live sports on 29, 45, 49–50, 87, 93, 161
    news on 34
    original content on 48–49
    subscriber numbers 162
  Twitch, purchase of 111
  Unbox 13
Anys, Imane (Pokimane) 115
Apple 6, 7, 33, 43–44, 50–55, 60–61, 90
  AI use 183
  App Store 61
  Apple Books 49, 137
  Apple Music 11, 128, 132–33, 138, 139, 140
    royalties paid 142, 143
  Apple One 52–53
  Apple TV 26, 62
  Apple TV+ 2, 6, 12, 51–52, 53–55, 62, 63, 72, 187
    binge watching on 190
    in China 154
    CODA 8–9, 54–55
    cost of 169
    live sports on 9, 29, 33, 55
    original content on 2
    on Prime Video Channels 61–26
  iCloud 53
  iPhone 7, 51, 154–55, 181
  iPod 147–48
  iTunes 11, 13, 15, 129–31, 132, 136
  podcasts 12
  relationship with Disney 63–64

Siri 149
TV box 77
Vision Pro 64
Armer, Craig 139
artificial intelligence (AI) 176–86
  on Amazon 176–77
  on Apple 183
  defining 176
  for moderation 123
  Hollywood concerns over 2, 4–5
  jobs, threats to 185, 192
  on Netflix 177, 179–80, 181
  OpenAI 176, 183
    ChatGPT 4
    Sora 5, 178
  Runway 184
  'slop' 186
  on Spotify 182–83
  StabilityAI 184
  on YouTube 178
audiobooks 136, 137–38
Avengers: Endgame 75, 76

Baidu 155
Bajaria, Bela 32, 34, 96
'Barbenheimer' 75, 76
Barone, Mary Beth 119
Barrie, Danielle (Dando) 121
Batgirl 70, 192
Bauer 148–49
Bauer, Helen 119
BBC 95, 97, 158
  BBC Sounds 148
  BBC World News 156
  iPlayer 72–73, 85, 170
  licence fee 160, 170
Beatport 141
beIN Sports 91
Beyoncé 89
Bezos, Jeff 45, 49, 111
binge watching 7–8, 188–90
Blavatnik, Len 89–90
Blevins, Tyler (Ninja) 108
Blockbuster 16–17, 18, 197
Bloomberg 50
  Bloomberg TV+ 68
Bohart, Catherine 119
Brewster, Bill and Broughton, Frank 140
Brody, Adrien 177
Bronze, Lucy 72
Brutalist, The 177, 185
Burns, Michael 184
Bush, George W 147
Bytedance 155
Cameron, James 184
Carney, Brittany 119
CBC Gem 160
CBS Network 83
Cenat, Kai 108
censorship, in China 156–57
Chappelle, Dave 36
Child, Lee 48
children's content 172–74
China Central Television (CCTV) 156
China Global Television Network (CGTN) 156–57
Christchurch massacre 122–24
cinema, impact of streaming on 73–77, 196–97
Clancy, Dan 111
Clancy, Tom 48
CNBC 68
CNN 50, 68
  CNN+ 71
Cocomelon 172–73
Comcast/NBC Universal 85
Cook, Tim 53, 55, 62, 154
Corrin, Emma 185
cost of living crisis 171
Covid-19 pandemic 7, 54
  cinemas, impact on 73–75, 196–97
  and livestreaming 118–20
  Play Something 179–80
  podcasts 12, 145
'Crown Jewels' 9–10
Cue, Eddy 132

Daily, The 148
Dall-E 4
Dalrymple, Adam 87
DAZN 89–92, 161
Deezer 139, 182
DeGeneres, Ellen 36
Disney 43, 62–66, 107
    Disney+ 2, 62–66, 105
        advertising on 10
        in Africa 157
        AWS, use of 63
        in China 154
        cost of 168–69
        ITV, alliance with 64–66
        subscriber numbers 162
    Hotstar 152
    relationship with Apple 63–64
Distard, Pete 103
Donaldson, Jimmy (MrBeast) 57–59
Douyin 155

Eacott, Lannan (LazarBeam) 113
Earley, Joe 65
EHFTV 97–98, 168
Ek, Daniel 12, 133–34
Emilia Perez 177
EmilyCC 115–16
Erlicht, Jamie 12, 51, 55
ESPN
    DTC offering 104, 105, 194
    ESPN+ 9, 103
    Venu 103–05

Fanning, Shawn 128
FA Player 95–96
Flavall, Stephen (jorbs) 114–15
Fox
    Fox News 59
    Tubi 72
    Venu 103–05
Foxtel 161
Frank, Michael 82, 92, 98

free ad-supported television (FAST) 67–68
    ITVX 65, 73, 160
    Samsung TV+ 67–68, 72
    Tubi 72
Freesat 157
Fubo 104–05, 161, 194
Fukunaga, Cary 28

Gadsby, Hannah 36
gamepass model 102
gaming
    burnout 114–15
    and misogyny 121–22
    on Netflix 39–41
    and terrorism 123
    top livestreamers 108–09, 113–14, 115
    see also Twitch
Gigless 119
Global Networks 70
Global Radio 149
Goalhanger 148
Google 176
    AdSense 112–13
    see also YouTube
Gregoire, Dannie 145
Gupta, Sanjog 153–54

hallucinations 175
Hammersley, Ben 145–46
Hanks, Tom 75–76
Hanson, Scott 85
Harraghy, Daniel 82
Harrington, Tom 32, 45, 52, 53
Hasan, Mahdi 59
hashing 125
Hastings, Reed 12, 15, 20–21, 37, 38–39, 40–41
HBO 5, 157
    Max 70, 162, 163, 168
    Now 7
H.E.R 120
Hirsch, Jeffrey 187

Index

Hofstetter, Rachell (Valkyrae) 113–14
Hollywood strikes, 2023 1–2, 3–4, 41, 191–92
Holmes, Karl 66
Hopkins, Mike 47
House of Cards 5–6, 19
How I Built This 137
Hulu 157

IBM 176
Iger, Bob 63–64
Infantino, Gianni 96–97
Instagram 56, 109, 110, 117, 196, 198
   in China 155
   Instagram Live 119–20
iQYI 155, 162
iROKTOW 158
ITV
   Disney+, alliance with 64–66
   ITVX 65, 73, 160

Jassy, Andy 176–77
JioHotstar 152–54
Jobs, Steve 11, 64, 129–30, 133, 134
John, Elton 120, 132

Kan, Justin 112
Katz, Don 137
Kayo 84, 161
Kelly, Megyn 59
Kiehl, Tom 130, 131, 138, 139, 140, 143, 144
Koenig, Sarah 146
Korzen, Jonathan 146

Lanternier, Alexis 182
Last Night a DJ Saved My Life 140
Lawrence, Bill 54
Legend, John 120
Leitch, Shirley 123
Lellobee City Farm 173

Lemon, Don 59
Lennox, Annie 142
LFP 91
LimeWire 129
Lionsgate
   Lionsgate+ 71, 187
   Runway, deal with 183–84
Little Baby Bum 173
Lorentzon, Martin 132
Lowe, Zane 132–33
Lygo, Kevin 65

Mango TV 155
'marathons' 110
Marchand, Andrew 104
Martin, Chris 120, 142
Ma, Wayne 154
McCarthy, Barry 16
McCartney, Paul 142
McHenry, Jackson 188
McIlroy, Rory 81, 98
Meta 183
Meyer, Erin 38
MGM
   MGM+ 50
   purchase by Amazon 46–48
Microsoft 24, 176, 183
   Teams 118
Midjourney 4
mobile tech 7
moderation, of social media 124, 125
Moonbug 172, 173
Moore, Danni 81
Morgan, Piers 59
MrBeast 198
Murdoch, James 152
Murdoch, Lachlan 72

Napster 11, 127–29, 130, 133
NBC 59
   Peacock 65, 71–72, 168
   live sports on 29, 83, 84–85
   advertising on 85

Netflix 2, 3, 9, 12, 13–42, 43, 54, 64, 72, 197
  advertising on 10, 22–25, 28, 32, 86
  in Africa 157, 159
  AI use 177, 179–80, 181
  algorithm 26–28
  binge watching on 11, 189
  Blockbuster, attempt to buy 16–17, 18, 197
  children's content on 174
  in China 154
  company culture 37–39
    'keeper test' 38, 39
    'no rules rules' 37
  election night broadcast 34
  games on 39–41
  live sports on 29–33, 86, 88, 89, 96–97
  minutes of content viewed 109
  Netflix Cup 31
  Netflix is a Joke festival 34
  'Netflix model', the 14
  Netflix Slam 31
  news on 34
  original content on 5–6, 18–20
  password sharing 20–22, 169
  Play Something 179–80, 182
  'pretty okay TV' 187–88
  stand-up comedy on 35–37
  story of 15–22
  subscriber numbers 162
  TF1 deal 25–26
  vs tech firms 68
news 8, 34, 50, 68
Nollywood 158–59

Obama, Barack 183
Ofcom 156
O'Leary, Michael 197
OnlyFans 115
OpenAI 176, 183
  ChatGPT 4
  Sora 5, 178

Optus Sport 83–84, 161
Orange is the New Black 19–20

Pandora 149
Paramount 2
  Paramount+ 61, 83, 168, 169
Parker, Sean 128
Paul, Jake 88, 96
Peppa Pig 173–74
Pitaro, James 105
Pixar 63
Pluto 67
podcasts 12, 135–36, 145–48, 150
  advertising on 23
Premier Sports 50

Queen Elizabeth II 8

Randolph, Marc 12, 15, 16–17
Reacher 48, 50
RealPlayer 146
'rebundling' 103–05, 193–94
Red Notice 41, 75
Reinhard, Amy 25
Rethinking Social Media and Extremism 123
Richardson, Eileen 128
Rife, Matt 35
Rock, Chris 36
Rogan, Joe 136, 148
Ronaldo, Cristiano 91
rugby 84
Runway 184

Samsung 77
  Samsung TV+ 67–68, 72
Sarandos, Ted 27, 29, 30, 33
Segev, Shay 90
Seinfeld, Jerry 36
Serial 146–47
Serrano, Amanda 88, 96
Shah, Mihir 152
Showmax 157–58
Silverman, Sarah 183

SiriusXM 149
Sky 65, 94, 100, 156–57
   Lionsgate+ 187
   Now 93, 94, 99
   Sky Sports 9, 29, 79, 80, 93, 95, 97
Sky Sport (NZ) 161
SmartLess Media 148
Sonantic 183
Sony 129–30
Sony Pictures Television 51
SoundCloud 140, 141
sports, live 194
   and advertising 32, 81–83, 85–86, 170
   baseball 29, 55, 83, 102
   basketball 84, 102
   boxing 33, 88, 89–90
   cricket 99, 151–53, 161
   football (soccer) 9, 29, 49, 79, 80, 81, 83–84, 87, 91–92
      in Africa 158
      in Australia, New Zealand and Canada 161
      'black-out' 101–02
      fragmented access to 93–94
      Major League Soccer 55
      women's 91, 95–97
      World Cup 8, 9
   football, American 29, 32, 33, 49, 60, 89, 82–83, 84, 87
      Christmas Day 2024 games 89
      fragmented access to 94
      NFL Gamepass 91, 102
      Super Bowl 9
   football, Aussie rules 90
   Formula 1 racing 31
   golf 81, 98
      Netflix Cup 31, 32
   handball 97–98, 168
   horse racing 10
   Olympic Games 84–85, 155
   rebundling of 103–05
   rugby league 84, 90–91
   and subscriptions 83–85
   table tennis 155–56
   tennis 32, 49, 84, 103
      Netflix Slam 31
      Wimbledon finals 10
   up-and-coming sports 98
   VPN use 99–101
   wrestling 31–32, 33
Spotify 11, 15, 133–37, 138, 139
   advertising on 134
   AI use 182–83
   audiobooks 136
   podcasts 12, 135–36
   royalties paid 142–43
   video content 135, 136–37
Squid Games 21
StabilityAI 184
Stan 84
   Stan Sport 161
stand-up comedy 35–37, 119
Stern, Howard 149
Stokel-Walker, Chris 57
Stoker, Ryan 125–26
Stranger Things 11, 21, 41
'subathons' 110
Substack 59
Swift, Taylor 143

TalkTV 59
Taylor, Katie 88, 96
Ted Lasso 53–54, 187
Tencent Video 155, 162
terrorism 122–25
Thompson, Mark 71
Thunderflix 167–68
TikTok 35, 56, 109, 110, 117, 198
   in China 155
TNT 93, 94
   TNT Sports 80

Tortured Poets Department, The 143
Trump, Donald 8, 154
TuneIn 149
TVNZ+ 160
Twitch 107, 110, 111–12, 115, 118
  abuse / harassment on 121–22
  advertising on 113
  Halle terrorist attack 122, 123, 124
  minutes of content viewed 109
  user age 109
Tyson, Mike 88, 96

UK Cinema Association 73
Van Amburg, Zack 12, 51
Venu 194
Vilahamn, Robert 96
Virtual Private Networks (VPNs) 99–101, 160
Warner Music 129–30
Warner Bros Discovery 69–71, 103
  CNN 50, 68
    CNN+ 71
  Discovery+ 71, 93, 94
  HBO Max 70, 162, 163, 168
Watkins, Elliott (Muselk) 113
Wednesday 41, 187

Wilde, Jeff 46
William, Prince of Wales 101
Willow TV 161
Winfrey, Oprah 50–51
Wong, Ali 36
Woods, Tiger 98

Yangshipin 155
Youku 155
YouTube 55–60, 107, 117, 118, 198
  advertising on 23, 112–13
  AI use 178
  algorithm 57
  children's content on 172, 173, 174
  in China 155
  influencers on 57–58
  live sports on 95–96, 97, 98
  Partner Program 56
  podcasts 150, 188
  YouTube Live 108, 111, 112–13
  YouTube Music 139
  YouTube TV 60, 104
    live sports on 9, 29, 60
YouTubers 57

Zaslav, David 70
Zoom 118–19, 120

Looking for another book?

Explore our award-winning books from global business experts in General Business

Scan the code to browse

www.koganpage.com/general-business

Also available from Kogan Page

ISBN: 9781398600683

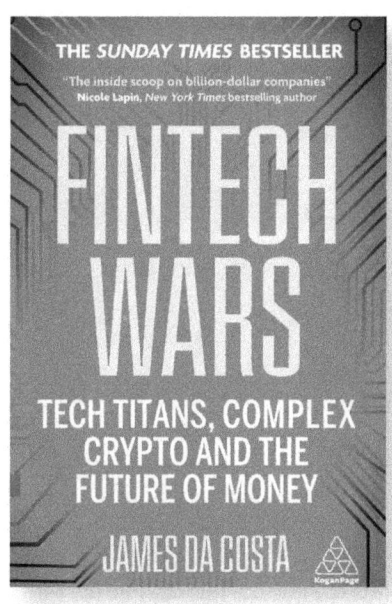

ISBN: 9781398617025

www.koganpage.com

From 4 December 2025 the EU Responsible Person (GPSR) is:
eucomply oÜ, Pärnu mnt. 139b – 14, 11317 Tallinn, Estonia
www.eucompliancepartner.com

www.ingramcontent.com/pod-product-compliance
Lightning Source LLC
Chambersburg PA
CBHW051534020426
42333CB00016B/1919